Scientific
Computing on
Supercomputers II

Scientific Computing on Supercomputers II

Edited by

Jozef T. Devreese and
Piet E. Van Camp

University of Antwerp
Antwerp, Belgium

PLENUM PRESS • NEW YORK AND LONDON

Library of Congress Cataloging-in-Publication Data

International Workshop on the Use of Supercomputers in Theoretical
  Science (5th : 1989 : University of Antwerp)
    Scientific computing on supercomputers II / edited by Jozef T.
  Devreese and Piet E. Van Camp.
      p.   cm.
    "Proceedings of the Fifth International Workshop on the Use of
  Supercomputers in Theoretical Science, held November 29-30, 1989, at
  the University of Antwerp, Antwerp, Belgium"--
    Includes bibliographical references and indexes.
    ISBN-13: 978-1-4612-7914-3
    1. Supercomputers--Congresses. 2. Science--Data processing-
  -Congresses.   I. Devreese, J. T. (Jozef T.) II. Van Camp, P. E.
  (Piet E.) III. Title.
  QA76.5.I623 1989a
  004.1'1--dc20   90-7899
                                                               CIP

Proceedings of the Fifth International Workshop on
the Use of Supercomputers in Theoretical Science,
held November 29–30, 1989, at the University of Antwerp,
Antwerp, Belgium

ISBN-13: 978-1-4612-7914-3   e-ISBN-13: 978-1-4613-0659-7
DOI: 10.1007/978-1-4613-0659-7

© 1990 Plenum Press, New York
Softcover reprint of the hardcover 1st edition 1990
A Division of Plenum Publishing Corporation
233 Spring Street, New York, N.Y. 10013

PREFACE

The International Workshop on "The Use of Supercomputers in Theoretical Science" took place on November 29 and 30, 1989 at the University of Antwerp (UIA), Antwerpen, Belgium. It was the fifth in a series of workshops, the first of which took place in 1984.

The principal aim of these workshops is to present the state-of-the-art in scientific large scale and high speed computation. Computational science has developed into a third methodology equally important now as its theoretical and experimental companions. Gradually academic researchers acquired access to a variety of supercomputers and as a consequence computational science has become a major tool for their work.

It is a pleasure to thank the Belgian National Science Foundation (NFWO-FNRS) and the Ministry of Scientific Affairs for sponsoring the workshop. It was organized both in the framework of the Third Cycle "Vectorization, Parallel Processing and Supercomputers" and the "Governemental Program in Information Technology". We also very much would like to thank the University of Antwerp (Universitaire Instelling Antwerpen - UIA) for financial and material support.

Special thanks are due to Mrs. H. Evans for the typing and editing of the manuscripts and for the preparation of the author and subject index.

J.T. Devreese   P.E. Van Camp
University of Antwerp

April 1990

CONTENTS

VECTORIZATION, OPTIMIZATION AND SUPERCOMPUTER ARCHITECTURE

Willi Schönauer and Hartmut Häfner

Rechenzentrum der Universität Karlsruhe, Postfach 6980, D-7500 Karlsruhe 1, F.R.G.

ABSTRACT

The basic architecture of vector computers is discussed in general and in more detail for some relevant vector computers. The relationship between arithmetic operations, memory bandwidth and realistic performance is pointed out. The key for an optimal vector computer algorithm is the data structure. Some basic problems are discussed as examples for the selection of optimal data structures and algorithms. The basic principles and rules are extracted from these examples. Finally the design of software for vector computers independent of a special architecture is briefly mentioned.

1. THE ARCHITECTURE OF VECTOR COMPUTERS

Large scale computations, i.e. number crunching, always means a series production of numbers. So we can use the "assembly line" principle for this series production: the arithmethic pipeline or (vector) pipe. In Figure 1.1 a 5-stage addition pipeline is depicted: 5 operand pairs are in different stages of execution in the pipeline, thus the operations must be independent. The following Fortran loop

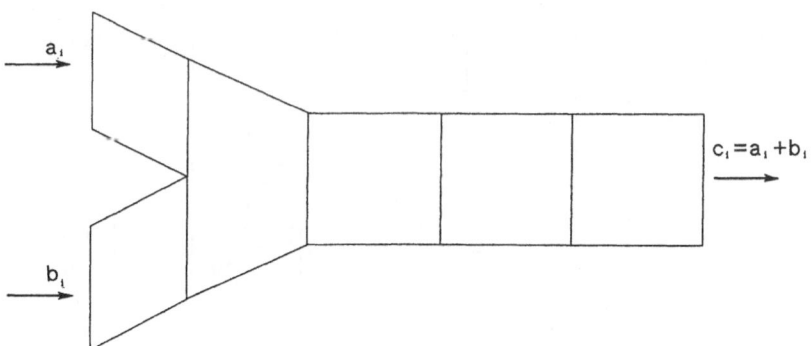

Figure 1.1. 5-stage addition pipeline.

Scientific Computing on Supercomputers II
Edited by J. T. Devreese and P. E. Van Camp
Plenum Press, New York, 1990

```
      do 10  i = 1,1000
          d(i) = a(i) + b(i) * c(i)                                    (1.1)
  10    continue
```

results in a compound vector operation in which the addition and multiplication pipes work in parallel. This is called supervector speed because two results are produced in each cycle time τ. But before the operation can start, the addresses of the operands and the vector length must be loaded to certain registers and (eventually) the operands must be loaded into vector registers. Then the pipes must be filled, the result address must be loaded to a register and the result stored from a vector register back to memory.

This situation leads to the following tuning formula for the processing of a vector of n elements:

$$t_n = \tau_{eff} (n + n_{1/2,eff}) \ .\tag{1.2}$$

The effective cycle time τ_{eff} is determined by the bottlenecks in the processing of the data and is (for a single pipeline) usually much larger than the hardware cycle time τ. The value $n_{1/2,eff}$ is Hockney's "half performance length" ([1], see also [2]), that represents the wasted (fictitious) operations for the startup of the vector operation under consideration. If $n = n_{1/2,eff}$, half the time is useful and half the time is wasted, thus only half of the effective peak performance is obtained. Another effect of $n_{1/2,eff}$ is, that if n is below some breakeven length n_b, it is cheaper to execute the n operations in scalar mode. Values $n_{1/2,eff}$ range between 20 (CRAY-1) and 200 (Fujitsu VP200) for simple operations, values of n_b are in the range of 4 to 10. The consequence of $n_{1/2,eff}$ for the programmer is the following

Rule: Make vectors as long as possible! (1.3)

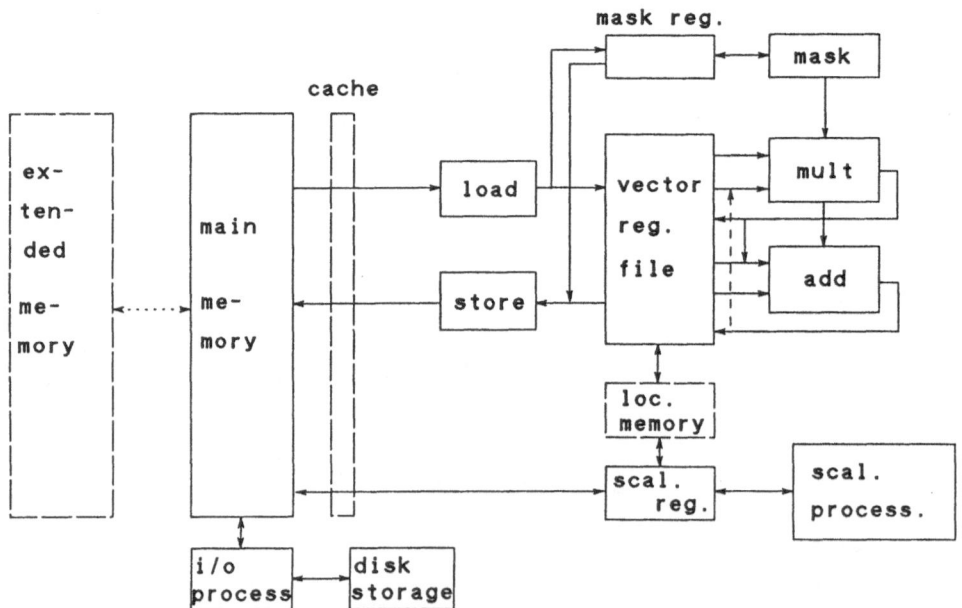

Figure 1.2. Overview of the main components of a vector computer.

For longer vectors the relative effect of $n_{1/2,eff}$ is smaller and one gets closer to the effective peak performance. An exception from this rule is the IBM Vector Facility (VF), where the locality of data plays an important role because of the pecularities of a cache that has been designed for general purpose and not for vector operations.

In Figure 1.2 an overview of the main components of a vector computer is depicted, dashed components are present only in special computers. Figure 1.2 represents a register vector computer in contrast to the memory-to-memory vector computers CYBER205 and ETA10 (see[2]), that are now history. The presence of vector registers of length N_R necessitates that a vector of length $n > N_R$ is processed in sections of length N_R and a remainder, i.e. the vector is strip-mined with length N_R and each section needs new startup cycles.

The essential bottleneck of all existing vector computers is the narrow memory bandwidth between the main memory and the vector registers. It is ultimately this bottleneck that limits the performance of the vector computers and not the performance of the pipes. We shall discuss this question in more detail below. Another weak point is the size of the main memory that practically limits the size of the problem that can be treated on the vector computer because disks are by far too slow compared to the processing speed of the pipes. An extended memory that is available for some vector computers can help, but here also the memory bandwidth usually is not sufficient. A more detailed discussion of architectures can be found in [1,2].

If we want to increase the performance of vector computers, there are basically four possibilities:
1. reduce the cycle time τ. This is limited by the available technology.
2. use pipelining. This is limited by the possibility to subdivide an operation.
3. use internally parallel pipelines. In this case the effective half performance length for N pipes is the N-fold of a single pipe:

$$n_{1/2,eff} = N * n_{1/2}.$$

Table 1.1. Characteristic data of several vector computers.

computer	cycle time [nsec]	no.of pipe groups per proc.	no of proc.	theor. peak perf. [MFLOPS]	words per cycle and pipe group	more realis- tic perf. [MFLOPS]	max. main mem. [Mwords]	max. extended mem. [Mwords]	length of vector register [words]
CRAY X-MP	8.5	1	4	941	3	706	16	512	64
CRAY Y-MP	6	1	8	2667	3	2000	128	512	64
CRAY-2	4.1	1	4	1951	1	488	512	-	64
Fujitsu VP200-EX	7	2x1.5	1	857	1.33	286	128	-	up to 1024
VP400-EX	7	4x1.5	1	1714	0.67	286	128	-	up to 2048
VP2000/600	4	8	1	4000	1	1000	256	1000	up to 2048
NEC SX-3	2.9	8	4	22069	0.5	2759	256	2000	256
IBM 3090/600S+VF	15	1	6	800	1	200	64	256	256
CONVEX C240	40	1	4	200	1	50	512	-	128

4. use parallel processors. This means that the problem of parallelism is shifted to the software level, i.e. to the user. There are three basic types of parallel processors:
   a) shared memory systems,
   b) message passing systems (distributed memory),
   c) hybrid systems (shared and local memory).

In table 1.1 characteristic data of 7 vector computers, of the IBM 3090 Vector Facility (VF) and of the CONVEX minisupercomputer have been compiled. The Fujitsu vector computers are marketed in Europe by Siemens as VP- or S-Series. The first data column indicates the technology and shows that the leading technology of supercomputers is five times faster than the technology of the large main frames like the IBM 3090 and 13 times faster than the off-the-shelf technology of minisupercomputers. The second and third data columns indicate the internal and external parallelism of pipes and processors. By "pipe group" we denote a group of addition and multiplication pipe. The value 1.5 for the VP-EXs means that there is a multifunctional pipe (addition and multiplication) and a separate addition pipe, i.e. 1.5 "groups".

The fourth and sixth data columns give the performance in MFLOPS (million floatingpoint operations per second). The theoretical peak performance results from the cycle time and the total number of pipes for supervector speed. But unfortunately the performance is limited by the memory bandwidth. A completely free usage of supervector speed would be possible only for the vector triad like in the loop (1.1). But this necessitates 4 words per cycle and pipe group (3 loads and 1 store) memory bandwidth. None of the existing vector computers has this memory bandwidth. Therefore we multiply the theoretical peak performance by the number of words per cycle and pipe group, indicated in the fifth data column, and divide by 4 (the necessary memory bandwidth). This is what we call a "more realistic performance". Many benchmarks that we have executed for the selection of supercomputers have demonstrated that this scaling gives an excellent performance value for engineering problems. The success of the CRAY X-MP and Y-MP is due to a large extent to the memory bandwidth of 3 words per cycle and pipe group, whereas the extremly low value of 0.5 of the NEC SX-3 reduces drastically the performance.

Data columns seven and eight give the size of the main and extended (where available) memory in 64-bit words. In the last data column the length of the vector register is presented. CRAY cannot get rid of its old value of 64, Fujitsu has a variable length of the vector registers (register file).

It is interesting to have a look at the machine instructions that are executed for a loop like (1.1) for the vector triad. For this purpose the Fortran source code and the assembler code for the processing of one section of a vector is presented below for the IBM VF and CRAY X-MP. The preparation and termination part of the programs are omitted. We present copies of the slides that have been used at the conference. Hand-written comments are inserted for explanations. These examples show that many (repeated) scalar operations are needed for the execution of a vector operation.

In Figure 1.3 the section loop for the IBM VF is presented, i.e. for the processing of 256 elements (section length) or a remainder. The VF has no chaining, i.e. overlapping of load and execution, but it has a "vector multiply and add" operation where the result of the multiplication pipe is immediately delivered to the addition pipe. One operand can be obtained directly from the cache.

Because of the memory bandwidth of one word per cycle between cache and vector unit, 4 chimes (time to process n elements) are needed, of which only one

```
        24          CALL SECOND(TCPU1)
        25          DO 60 R=1,M
1       26            CALL DUMMY(A,B,C)
1       27            DO 60 I=1,N
2       28              D(I)=A(I)+B(I)*C(I)
2       29     60    CONTINUE
```

} *vector triad loop*

(multiply and add operation)

VR = *vector register*

$$d_i = a_i + b_i * c_i$$

VR∅   VR∅   VR2   *cache*

c from cache

4 chimes: load a, load b, operation, store d

only "useful" chime

assembler code:

in reg.9: 8*N (8N Bytes = N words), in reg.10: 8 (8 Bytes = 1 word), in reg.8: 2048 (8·256)

```
0014FE  8E20 0020                   SRDA   2,32              8  gives N→3
001502  5D20 D0FC                   D      2,252(0,13)          N→6
001506  1863                        LR     6,3
001508  A645 0060                   VLVCU  6                 .00002  min({6},256)→VC
* ISN   28                                                           {6}-256→{6}
00150C  410A 7FA0     section  3.111 LA    0,4000(10,7)      B     .Z.000,B } b→VR2
001510  A419 0020     Loop for       VLD   2,0(0)            A     .Z.001,A } a→VR∅
001514  410A 7000     256 el.        LA    0,0(10,7)                         } multi. and
001518  A419 0000     or rest        VLD   0,0(0)            C     .Z.002,.Z.000,C } add→VR∅
00151C  410A 5000                    LA    0,0(10,5)
001520  A414 2000                    VMAD  0,2,0(0)          D     .Z.002..Z.000,C
001524  410A 4010                    LA    0,16(10,4)        D,..Z.003 } store d
001528  A41D 0000                    VSTD  0,0(0)
* ISN   29
00152C  A645 0060          4.111      VLVCU  6               .00002  update vector length
001530  87A8 B0B2                     BXLE   10,8,178(11)    3.111# →{reg.8+reg10→reg10,
001534  58B0 C054          27.001     L      11,84(0,12)     27.001# compare reg.10 with
001538  41A0 0008                     LA     10,8            8        reg.(8+1=)9,
                                                                      if {10} le {9} go to 3.111}
```

this is "one" (compound) vector instruction!

Figure 1.3. Source code and assembler code of the vector triad for the IBM VF.

chime is useful. So the theoretical performance is reduced by a factor 1/4. A further reduction results from the startups of the vector operations and from the execution of the auxiliary scalar operations. The details of the processing can be seen from the inserted comments.

In Figure 1.4 the section loop for the CRAY X-MP for processing 2 sections of 64 elements is presented. This needs an explanation. This loop is preceded by an eventual section to process the remainder of n mod(64) elements and an eventual section of 64 elements. The CRAY X-MP has chaining, i.e. overlapping of load/store and execution. The vector triad needs three loads and one store, but the X-MP has only two loads and one store pipe. Therefore we need two chimes (time to process 64 elements) for the processing of 64 elements of a vector triad: In the first chime we use two loads and loose the store, in the second chime we use one load and the store and loose one load, the multiplication and addition are overlapped with the loads (chaining). But if we execute two sections of 64 elements in an interleaved manner, we loose only one store. This is indicated in the upper part of Figure 1.4. In the part of the assembler code the processing of 64 elements and the boundaries of the chimes are indicated. All operations in one chime are executed in a "chained" form. If we had three loads and one store, we could process 64 elements in one chime. Therefore for the eventual two clearup loops the performance is reduced by a factor of 1/2 and for the interleaved processing of two sections by a factor of 2/3. This example illustrates >also the difficulties for correct timing formulas.

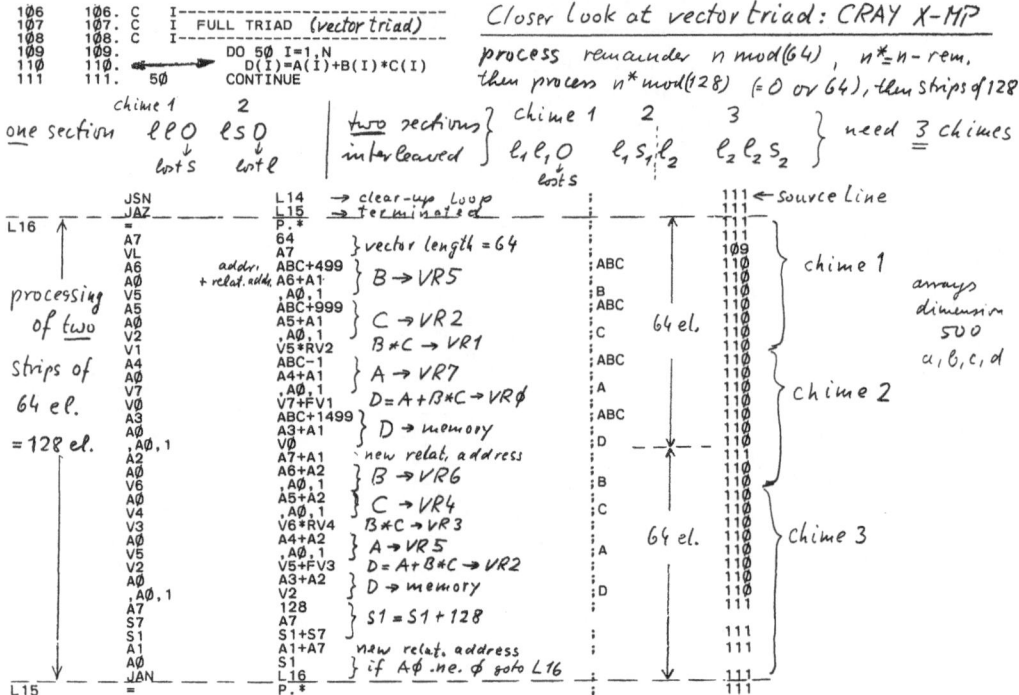

Figure 1.4. Source code and assembler code of the vector triad for the CRAY X-MP.

2. ARITHMETIC OPERATIONS, MEMORY BANDWIDTH AND MEMORY ACCESS

Because of the memory bottleneck, indicated in Table 1.1 by "words per cycle and pipe group", we must distinguish the operations by the number of necessary memory references.

Dyadic operations:

$$c_i := a_i + b_i \, , \quad c_i := a_i * b_i \, , \quad c_i := a_i / b_i \tag{2.1}$$

need 2 loads and 1 store (per cycle and pipe group, this is not explicity mentioned below). The divide may be much slower than addition or multiplication, then it needs correspondingly less memory references.

Triadic operations:

$$d_i := a_i + b_i * c_i \, . \tag{2.2}$$

This is the vector triad, it needs 3 loads and 1 store. The vector triad is the most important operation and gives the greatest flexibility for supervector speed, i.e. parallel operation of the addition and multiplication pipe. The following operations are special cases of the vector triad.

$$c_i := a_i + s * b_i. \tag{2.3}$$

This is the linked triad, s is a scalar. It needs 2 loads and 1 store.

$$c_i := c_i + a_i * b_i \text{ , repeated.} \tag{2.4}$$

This we call the contracting vector triad. It is the dominant operation for the simultaneous solution of many (small) linear systems if we vectorize over the number of systems. If we succeed to fix c_i in a vector register, we need for the repeated execution 2 loads, neglecting initial load and final store of c_i.

$$b_i := b_i + s * a_i, \text{ repeated.} \tag{2.5}$$

This we call the contracting linked triad, often denoted as SAXPY. It is the dominant operation for the solution of a linear system or of the matrix multiplication for full matrices. If we succeed to fix b_i in a vector register, we need for the repeated execution only 1 load, neglecting the initial load and final store of b_i.

A special triadic operation is the scalar product or dotproduct

$$s := s + a_i * b_i, \tag{2.6}$$

that needs 2 loads.

For dyadic operations we can get only vector speed. For triadic operations we can get supervector speed, doubling the performance. But if the performance is determined by the memory bottleneck we must know exactly the type of operation. In this sense the operation counts in most books on numerical analysis are obsolete, counting only additions and multiplications. Between a contracting linked triad and a full triad we have a factor of 4 in performance on a vector computer with a memory bottleneck of 1 word per cycle and pipe group. Such a computer should be called a "SAXPY computer" because its full performance can be obtained only for the contracting linked triad, i.e. the SAXPY. Only a vector computer with a memory bandwidth of 4 words per cycle and pipe group has a real chance to come with its sustained rate close to the theoretical peak performance. If the manufacturers "demonstrate" the performance by the matrix multiplication, they demonstrate the performance for an exception, i.e. for the contracting linked triad, and not for the general case. Wonder what NEC will demonstrate for its SX-3 with 0.5 words per cycle and pipe group.

If we want to characterize the realistic performance of a supercomputer we should use the following formula:

$$r_{real} = \frac{1\,000}{\tau\,[\text{nsec}]} * 2 * P * \frac{m}{4} * d * f\ [\text{MFLOPS}]. \tag{2.7}$$

The first factor gives the theoretical peak performance in MFLOPS of a single pipe for vector speed if we count the cycle time τ in nsec. The factor 2 is for supervector speed, i.e. for triadic operations, the factor P is the number of pipe groups for internal and external parallelism. Up to here we have what is called the theoretical peak performance. But this assumes triadic operations and thus we scale the performance by the relation m/4, with m the memory bandwidth in words per cycle and pipe group, the reason now should be clear. For a mere "SAXPY-machine", that has to compute only contracting linked triads, instead of m/4 the value 1 could be used for m≥1 and m for m≤1. Up to here we have what is called in Table 1.1 "more realistic performance".

The factor

$$d = \frac{n}{n + n_{1/2,eff}} = \frac{1}{1 + \frac{n_{1/2,eff}}{n}} \qquad (2.8)$$

accounts for the influence of the finite vector length. Here n and $n_{1/2,eff}$ are the "mean" values of a whole program for the vector length n and half performance length $n_{1/2,eff}$. The value of d depends on the architecture of the vector computer by the $n_{1/2,eff}$, but also on the program to be executed by the value of n. Thus the manufacturer <u>and</u> the user are responsible for this factor d. The user may see from the value of d how the performance changes if he changes the "dimension" n of his problem.

The factor f denotes the fraction of time that the pipes are busy, excluded the waiting of the pipes caused by the memory bottleneck that is accounted for by m/4. We have $0 < f \leq 1$. The manufacturer is responsible for the waiting of the pipes caused by large bank-busy-time and coarse memory structure, by memory contention and synchronization for multiprocessor systems. The user is responsible for the waiting of the pipes caused by scalar code, by bank conflicts and bad data structure, and by bad load balancing for multiprocessor systems. A special case are message passing (distributed memory) systems where the communication between the processors plays a decisive role: The manufacturer is responsible for the structure and speed of the communication system, the user is responsible for the ratio of communication to computation by the choice of his data structure and algorithm.

Thus the formula (2.7) makes clear why the real performance may be so far below the theoretical peak performance. The reducing factors are clearly visible and the user, but above all the manufacturer, can see the weak points of a computer for a certain problem. But the performance that is lost by a poor architecture can never be regained by the user, even by using the best algorithm.

The user has also to care for an optimal use of the memory hierarchy, e.g. he must avoid unnecessary load/store of vector registers by his algorithm or even by unrolling of other loops (see below). He never should expect that the compiler cares for all these items, such a compiler does not exist yet. For the IBM VF and in certain cases also for the CONVEX the user has to care also for data locality for an optimal use of the cache to avoid cache rolling.

An important point is the type of memory access. There are three basic types of memory access, namely
1. by contiguous elements: This is the optimal and only efficient way of memory access because it means a continuous "streaming" of the data.
2. by constant stride (e.g. every 2nd element is stride 2): This introduces the danger of bank conflicts and reduces the transfer rate for most computers. For multiprocessor systems the danger of memory contention is increased and for systems with virtual memory the use of the memory pages and above all of the cache is correspondingly reduced. If we have N banks and stride N all the data are in the same bank and for each element one has a waiting time that corresponds to the bank-busy-time.
3. by index vector, indirect addressing:

$$a(k) = b(i(k)), \text{ gather;} \qquad a(i(k)) = b(k), \text{ scatter.} \qquad (2.9)$$

This means increased danger of bank conflicts and, depending on the hardware, reduced access rate because also the index vectors must be loaded into a vector register. The early CRAY computers had no hardware instructions for indirect addressing, therefore such operations were executed in scalar mode. All types of problems with unstructured data depend strongly on gather/scatter operations.

For the mapping of multidimensional arrays to the linear addressing of a memory we have a fatal language dependency. The 4x4 matrix $A = (a_{ik})$

$$A = \begin{bmatrix} a_{11} & a_{12} & a_{13} & a_{14} \\ a_{21} & a_{22} & a_{23} & a_{24} \\ a_{31} & a_{32} & a_{33} & a_{34} \\ a_{41} & a_{42} & a_{43} & a_{44} \end{bmatrix} \qquad (2.10)$$

is mapped to the memory in Fortran by columns and in Pascal by rows. If we look at the matrix (2.10) we see that in Fortran/Pascal the elements are in contiguous locations if the first/last index is running and we have stride 4 (in general stride n for a nxn matrix) if we access the matrix by rows/columns. Therefore we have for the processing of multidimensional arrays by contiguous elements the following

Rule: In Fortran (Pascal) the innermost loop must run over the first (last) index! (2.11)

In the context of memory access the granularity of the memory plays a central role, i.e. the number of memory banks and the bank-busy-time (bbt).

A bad example is the CRAY-2 installed at the University of Stuttgart, one of the early models. The memory has 256 Mwords (million words of 64 bits), subdivided into 4 quadrants with 32 banks, each. Thus we have a total of 128 banks that behave like 256 banks for odd/even access that includes access by contiguous elements. The bbt is 61 cycles (for odd/even access the memory behaves like 128 banks with 37 cycles bbt). Because each of the 4 processors can access a quadrant only every 4th cycle we have the danger of quadrant and bank conflicts. The memory bandwidth for each processor is one word per cycle. Thus we need for general access 61 * 4 = 244 banks, but there are only 128 banks. Only for odd/even access by all 4 processors we have 256 banks available which meets the requirements. As soon as only one of the 4 processors uses non-contiguous access it will severely disturb the 3 other processors. This effect can be seen by the extremely wide scattering of measurements for the CRAY-2, see [2]. In the meantime CRAY has improved and is still continuously improving the granularity of the memory of the CRAY-2, but nevertheless this bad example demonstrates drastically the reason for memory contention in multiprocessor systems.

An example of an excellent granularity of the memory is the CRAY Y-MP with 8 processors and a memory of 32 Mwords, subdivided into 256 banks, the bbt is 5 cycles. Each processor has a memory bandwidth of 3 words per cycle for arithmetic operations (2 loads, 1 store) and 1 word for I/O. So without/with I/O we need 8 * 3 * 5 = 120 / 8 * 4 * 5 = 160 banks, compared to the available 256 banks. The consequence is that there is only little memory contention by the 8 processors.

These examples demonstrate that a computer with coarse granularity of the memory is much more susceptible to non-contiguous memory access, above all in a multiprocessor environment. Therefore the type of memory access plays a central role in the development of vector computer software.

3. DATA STRUCTURES AND THE DESIGN OF ALGORITHMS

Let us assume we had a program for a certain problem and a vector computer with a scalar performance $r_{eff,scal}$ (e.g. in MFLOPS) and a vector performance $r_{eff}(\bar{n})$ for a "mean" vector length $\bar{n}$ of that program. In the program the part v of the arithmetic operations is vectorizable, thus 1-v is the scalar part. Then the speedup, going from scalar to vector operations (see [1,2]) is the relation of the time to execute the program in scalar mode to the time to execute the program in vector mode

$$sp(\bar{n},v) = \frac{t_{scal}}{t_{vector}} = \frac{1}{v \dfrac{r_{eff,scal}}{r_{eff}(\bar{n})} + 1 - v} \, . \tag{3.1}$$

This is called Amdahl's law. The formula (3.1) is rather academic, because we never know the entries for a large program with sufficient accuracy, but nevertheless we can see two interesting limiting cases of (3.1): For v = 1, 100% vectorizable code, we get

$$sp_{max}(\bar{n}) = r_{eff}(\bar{n})/r_{eff,scal} \, , \tag{3.2}$$

i.e. the ratio of the vector to scalar speed. For $r_{eff}(\bar{n}) = \infty$, infinite vector speed, we get

$$sp_{asymptotic}(v) = \frac{1}{1-v} \, . \tag{3.3}$$

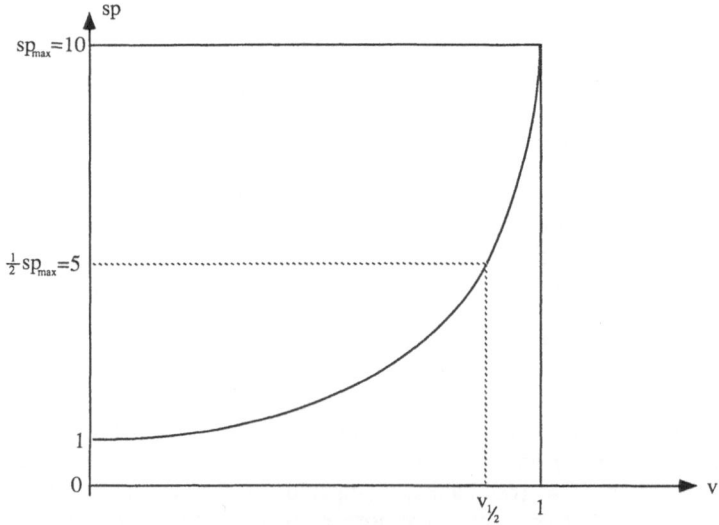

Figure 3.1. Speedup (3.1) for $r_{eff,scal}/r_{eff}(\bar{n}) = 1/10$.

This means that if we have v = 0.8 or 0.99 (80% or 99% vectorizable) we cannot get, even for infinite vector speed, a speedup larger than 5 or 100.

In Figure 3.1 the relation (3.1) is depicted for $r_{eff,scal}/r_{eff}(n) = 1/10$. Only for values of v close to 1 (100% vectorization) good speedup can be expected. In order to get only half the maximal speedup (3.2) we need in Figure 3.1 already $v_{1/2} = 0,889$, i.e. 88,9% vectorization. This behavior is less/more pronounced if we have a smaller/larger ratio of vector to scalar speed. The consequence of Figure 3.1 is the following

Rule: Try to make your program 100% vectorizable with long and contiguous
vectors.   (3.4)

The last part of the rule has the purpose to bring sp_{max} as high as possible.

The next problem is now, how to design programs so that they are nearly 100% vectorizable. The key to an excellent vector program is the data structure and only then the algorithm must be selected on that data structure. Here we have to obey the following

Basic Rule: Separation of the selection and of the processing of the data. (3.5)

This means that we have to choose for each (sub)task at first the optimal data structure or to establish this data structure by sorting and merging, and then to process the data. This demonstrates the necessity and importance of data transfer operations like the mask-controlled pack, unpack, merge and the indexvector-controlled gather/scatter operations. The data must be "prepared" for the assembly line principle of the pipes. We have to note the stupidity of the Fortran vectorizers in the context of the limitations of the Fortran language. They can never change data structures, they change only code.

In the following sections we shall demonstrate the application of the above mentioned rules for some selected examples. We shall try to extract from these examples typical working principles that can be easily extended to other problems. We assume Fortran rules.

4. MATRIX MULTIPLICATION AND RELATED PROBLEMS

The matrix multiplication C = A * B for full nxn matrices is discussed as an example of the choice of an optimal algorithm on a given data structure. In Figure 4.1 the multiplication scheme is depicted. In the "usual" scalar product form an element c_{ij} is computed as scalar product of a row of A and a column of B. But accessing a row of A means (in Fortran) access with stride n. A better algorithm is to compute simultaneously all elements of a column of C as a "vector", the corresponding Fortran program is depicted in Figure 4.1. The innermost loop runs over the first index and the corresponding operation is a contracting linked triad with contiguous elements. For a fixed data structure the selection of the optimal algorithm means the selection of one of the nested loops as innermost loop. Basically this could be done by the compiler, but up to now only a few compilers under restricted conditions can make such loop interchanges.

A vector longer than the length of a vector register is processed in sections, see Figure 4.1. Each section of a column of C is loaded from and stored back to the memory for each cycle of the k-loop. This means a tremendous wasting of load/store operations, above all for computers with a narrow memory bandwidth! The only thing

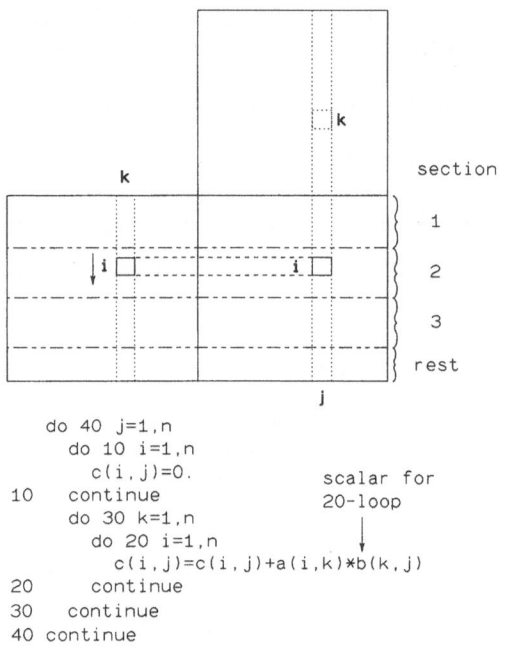

```
      do 40  j=1,n
        do 10  i=1,n
          c(i,j)=0.                scalar for
10      continue                   20-loop
        do 30  k=1,n                  |
          do 20  i=1,n                |
            c(i,j)=c(i,j)+a(i,k)*b(k,j)
20        continue
30      continue
40    continue
```

Figure 4.1. Illustration of matrix multiplication by columns.

that we can do on the Fortran level to reduce these load/store operations is unrolling of the next outer loop, in our case of the k-loop. For threefold unrolling (after a clear-up loop for n mod(3)) we execute instead of the loops 20,30 in Figure 4.1

```
      do 30 k = 1,n,3
      do 20 i = 1,n
      c(i,j)=c(i,j)+a(i,k)*b(k,j)+a(i,k+1)*b(k+1,j)
          +a(i,k+2)*b(k+2,j)                              (4.1)
20    continue
30    continue
```

This saves 2 loads and 2 stores for 3 contributions to c_{ij}. N-fold unrolling saves N-1 load/stores.

The optimal algorithm, that unfortunately cannot be expressed in Fortran, would be to "fix" a section of a column of C in a vector register and process all corresponding sections of A, thus terminating completely this section of C before going to the next section, see section 2 in Figure 4.1. Now we need only one load for a_{ik} in the loop 20 of Figure 4.1 as explained in the discussion of the operation (2.5). For the IBM VF further steps must be executed to meet the pecularities of a cache that has been designed for a general purpose computer and not for a vector computer. A "section stripe" of as many sections of A that fit into the cache is copied to a dense auxiliary array F in the memory (ultimately mapped into the cache) and then the contributions of this matrix F to the corresponding sections of all columns of C are computed, then the next section stripe of A is processed until finally the whole section of A has been processed. Then one proceeds to the next section. Here we see the priorities of the memory hierarchy: The fixing of F in the cache has highest priority, then the fixing of a section of C in a vector register during the processing of the contributions of F. Is this a user-friendly architecture?

The next problem that we want to discuss is the simultaneous multiplication of many small matrices as an example for the selection of an optimal data structure. Large computations result always from nested loops. Let us assume we have to compute m=10000 products of square nxn matrices, e.g. 10x10. This means we have a m-loop around the loops of the program of Figure 4.1. But it would be completely inefficient to vectorize like in Figure 4.1 over the innermost loop for n = 10. The trick is now to take the m-loop as the innermost loop. This means that we vectorize over all m matrices and thus compute the matrix products "simultaneously". But now we have to adapt the data structure correspondingly: We must store the elements of the matrices contiguously over the matrix numbers, see Figure 4.2. The ordering is to store the first elements of all matrices, then the second elements etc. For the matrix multiplication we can take any algorithm, e.g. the scalar product form or the columnwise form of Figure 4.1, because now these 3 loops are outer loops. The elements of the matrix A are stored as an array $a(l,i,k)$, where l runs over the number of matrices 1(1)m, similarly B and C are stored. We now have a l-loop around each operation as an innermost loop, as depicted in Figure 4.2. But now the operation is no longer a contracting linked triad (2.5) but a contracting vector triad (2.4) that needs two loads if we fix a section of the elements $c(l,i,j)$ in a vector register. Note that this section runs over part of the matrices, i.e. in l-direction.

At first sight this ordering over l seems to be unnatural. But usually the matrix elements are generated by a certain formula. If we generated each matrix individually, we had to generate the elements in scalar mode or with very inefficient vectorization over n. It is much more natural to generate the first element for all matrices, i.e. over m. This is just the data structure of Figure 4.2. The same arguments hold for the postprocessing of the matrix products. From this example we can extract the following

Rule: Select the optimal loop of a nest of loops for vectorization,
        adapt the data structure correspondingly. (4.2)

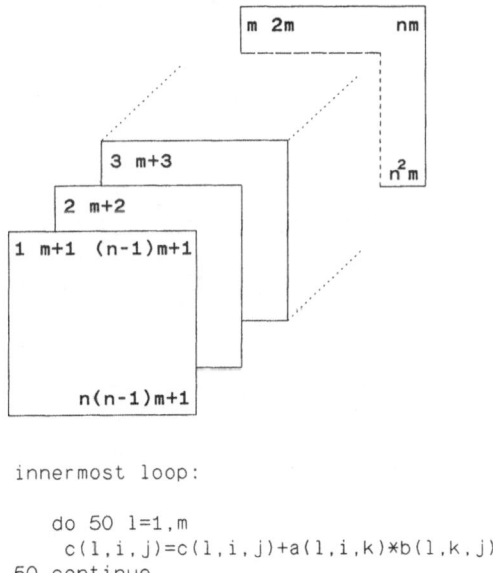

```
innermost loop:

   do 50 l=1,m
     c(l,i,j)=c(l,i,j)+a(l,i,k)*b(l,k,j)
   50 continue
```

Figure 4.2. Data structure for the simultaneous multiplication of many small matrices.

In the same way as we have discussed the matrix multiplication we can discuss the vectorization of the solution of a large linear system of equations or the simultaneous solution of many small linear systems, see [2]. The latter case happens e.g. for the generation of many different formulas on a body-oriented grid. But unfortunately these small linear systems are very critical and thus need pivoting during the elimination process. Pivoting means an individual treatment of each linear system and thus a simultaneous solution seems to be impossible. In Figure 4.3 the solution is depicted where for a Gauss elimination zeros are to be generated in the columns below the diagonal elements x. In two auxiliary vectors the absolute values of the diagonal elements and their row indices i are stored. Then in a vector compare operation the elements of the first vector are compared to the absolute values of the elements below the diagonal elements (row i=j+1) and where the latter are larger, the elements and their indices are exchanged. These are all operations over all m systems. This procedure is continued until the end of the pivot column. Finally in the index vector the indices of the pivot rows that must be exchanged with the diagonal rows are stored. For this interchanging the diagonal elements are stored to an auxiliary vector, the pivot elements are selected by a gather operation and stored to the diagonal position, and finally the elements of the auxiliary vector are stored by a scatter operation to the position of the selected elements. In the same way the elements of all following columns and of the right hand side of the linear systems are exchanged. Because most compilers cannot vectorize indirect addressing of intermediate indices one can compute the element positions in a "one-dimensional array" and use these indices that are now "shifted to the first index", for details see [2]. This example demonstrates how "individualism" can be treated by indirect addressing, i.e. by gather/scatter operations.

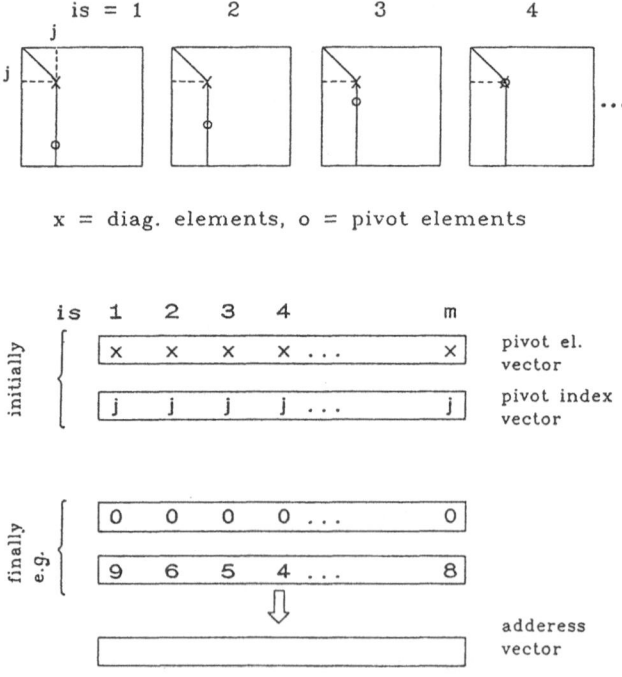

Figure 4.3. Illustration of pivoting for the simultaneous solution of many small linear systems.

5. RED-BLACK SOR AND DIAGONAL STORING OF MATRICES

We discuss red-black SOR as an example for the preparation of the data for an efficient vectorization. We assume that we want to solve an elliptic PDE (partial differential equation) with 2nd order central differences on a two-dimensional grid as depicted in Figure 5.1. We assume Dirichlet boundary conditions, i.e. given solution on the boundary. In each interior grid point a difference star couples 5 unknown values. For a SOR-type iterative solution of the resulting linear system we resolve each coupling equation for the central unknown (here we discuss only the questions related to vectorization). If we now number the unknowns in a red-black colouring, see Figure 5.1, the system decouples: red values are iterated from black ones and vice versa. Note that on the grid we can "count through" red, black, red, black,... also going from one row of the grid to the next one, but this needs the number nx of grid lines in x-direction to be odd. At this point the preparation of the data already starts.

We can see that we have the same index pattern for all interior points as depicted in Figure 5.1 a,b. But we have individual formulas for the grid lines adjacent to a boundary and for the corner points. Thus we could vectorize the interior grid points only along a single row of the grid. Could we not "prepare" the data that we get a single formula that vectorizes over all red grid points (the same then holds for the black ones)? The solution of this problem is as follows. We "process" the boundary points by multiplying the boundary values by their coefficients α and subtracting the result from the right hand side of the equations. Then these coefficients α are replaced by zeros. Further we have to extend the vector ub of the black values to the left and to the right by entier(nx/2) zeros. The resulting data structure is depicted in Figure 5.2. Now we can compute the red values by a single formula that holds equally for corner, boundary and interior stars and goes through all red points. The price that we have to pay are some dummy operations with artificial zeros. But one dummy operation is just one lost operation, a new vector operation costs $n_{1/2,\text{eff}}$ lost operations for the startup. Thus we have the following

Rule: Sacrifice dummy operations to avoid individual treatment.   (5.1)

A discussion of the extension of this principle to more general problems can be found in [2].

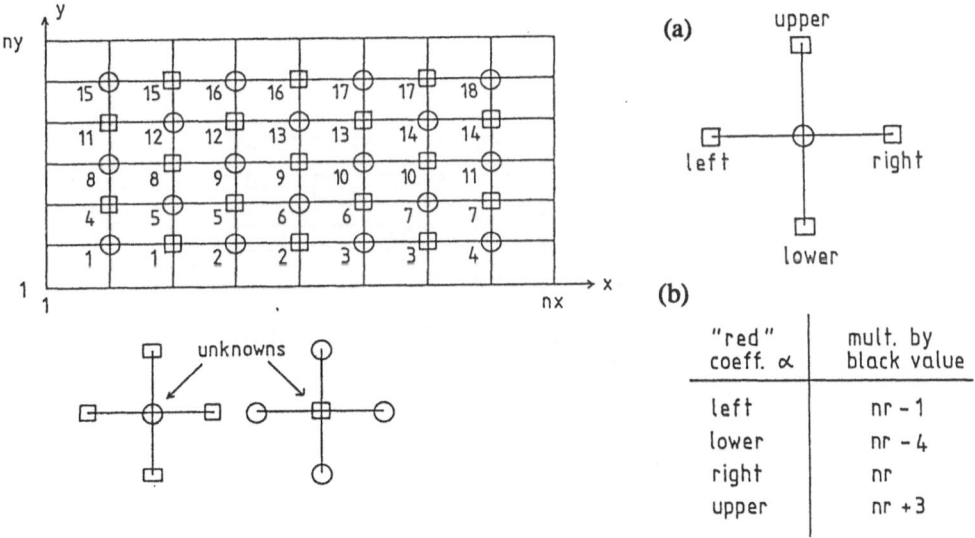

Figure 5.1. Difference grid and difference stars for red-black ordering.

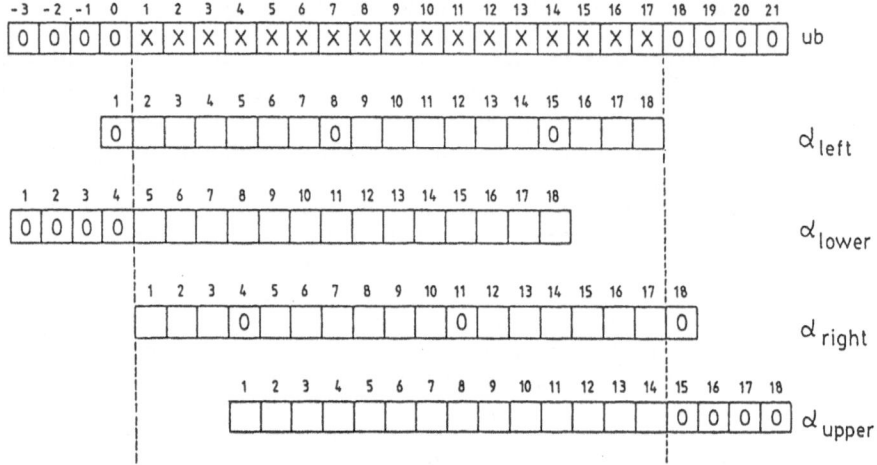

Figure 5.2. Data structure for the red formula.

An example for the search of an optimal data structure for an optimal algorithm is discussed now for packed diagonal storing. For finite difference methods (and to a certain extent also for finite element methods) the resulting matrix of the linear system of equations is very sparse, but it has the structure of scattered diagonals. Such matrices are stored as packed diagonals because the diagonals themselves are sparse. For each nonzero diagonal the nonzero elements and their indices in the diagonal are stored. Such linear systems are solved iteratively. The basic operation of all iterative solvers is the matrix-vector multiplication (MVM) that now must be formulated in diagonal form. The basic principle is depicted in Figure 5.3.

For the computation of A * r the upper/lower diagonals are arranged that the elements of the diagonals end/start with the vector r, the corresponding elements are multiplied and the products are added to the result vector A * r that the addition starts/ends with A * r. Because the diagonals are stored in packed form, there are 4 possibilities to process the diagonals: (1) to unpack the diagonals and process; (2)

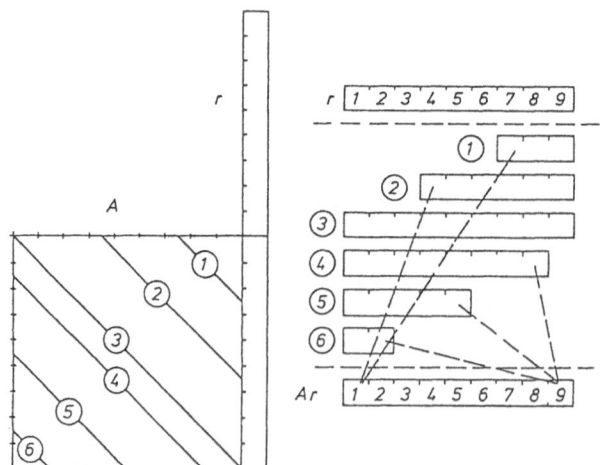

Figure 5.3. Matrix-vector multiplication in diagonal form, basic principle.

unpacking with unrolling; (3) unpacking with sectioning (for (2) and (3) there is the problem of the different length of the diagonals that can be solved by extending the shorter diagonals of a group by rule (5.1)); (4) to use sparse arithmetic with index vector (selection of the active elements). For the realistic examples that have been investigated, method (4) proved to be the best one and it was 2-4 times faster than method (1), see [3]. Clearly method (4) depends on the availability and speed of gather/scatter operations.

6. THE LINEAR FIRST ORDER RECURRENCE

The linear first order recurrence is discussed as an example how a seemingly non-vectorizable problem can be "vectorized" by the principle of "divide and conquer", but the price is a significant increase of arithmetic operations.

The linear first order recurrence is

$$x_i = a_i\, x_{i-1} + b_i \qquad\qquad (6.1)$$

with some starting value x_0. Such a sequence cannot be computed in a vector pipeline because x_{i-1} must be available before it can be used for the computation of x_i, so both cannot be in a pipeline at the same time. Introducing the relation for x_{i-1} into (6.1) yields

$$x_i = a_i^{(1)}\, x_{i-2} + b_i^{(1)}, \qquad\qquad (6.2)$$

combining x_i with x_{i-2}. Thus e.g. all odd or even x_i can be eliminated. The same method can be applied to the reduced sequence and so on until a single term remains. Then a backward substitution phase yields the intermediate elements of the preceding stage. This is the cyclic reduction, see [1,2]. The elimination and backward substitution are independent operations, but they are executed with stride or they need presorting. But the real price is that we need roughly 2.5 times the number of the arithmetic operations that are needed for scalar computation. Thus the vector operations must be more than 2.5 times faster than the scalar operations. This holds for practically all vector computers if the vector length is sufficiently large. A discussion for more general recurrences can be found in [2].

Another approach for the "vectorization" of (6.1) is the partition method, see[4]. Relation (6.1) corresponds to the solution of a bidiagonal linear system with a main diagonal "1" and an adjacent lower diagonal "-b_i", see Figure 6.1. This system is subdivided into equal parts (a remainder can be treated in scalar mode) and these parts are "disconnected", see Figure 6.1. The "carry" that links two adjacent parts is treated separately with a value "1". Then for all parts the recurrences are treated "simultaneously" in a similar way like the simultaneous multiplication of many small matrices in section 4. The vector length is the number of partitions. Now in a recursion the carries can be computed one after the other and in a final simultaneous computation over all parts the parts are completed by superposition of the precomputed "local" sequence and the influence sequence of the carry. By presorting the elements "over the parts" similarly to Figure 4.2 all computations can be executed with contiguous elements. Again the price is that we need roughly 2.5 times the operations of scalar computation, like for cyclic reduction. In [4] several variants of the partition method are presented. It would be an interesting task to compare the performance of the cyclic reduction to that of the partition method for different vector computers.

In real problems usually large computations result from the computation of many recurrences of all types (linear, nonlinear), e.g. for the numerical solution of large

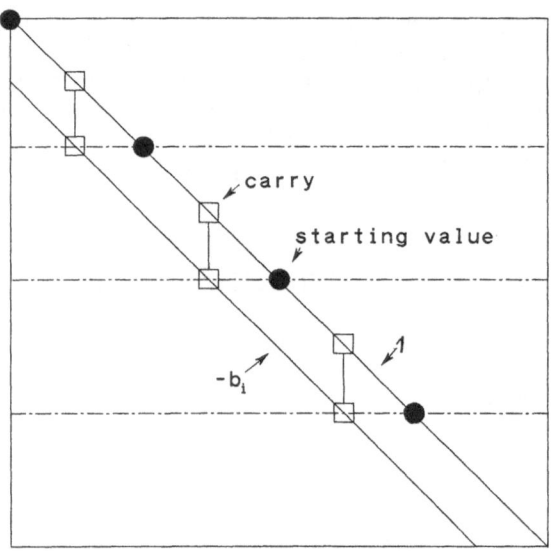

Figure 6.1. Illustration of the partition method.

systems of ordinary differential equations as an initial value problem (that result e.g. from the method of lines for parabolic PDEs). Then clearly the vectorization over the number of recurrences is to be preferred and again the data structure must be "over the systems" like in Figure 4.2. Then there are no extra operations.

7. GENERATION OF RANDOM NUMBERS

In this section we discuss the generation of random numbers purely from the point of view of vectorization. We do not discuss properties of different random number generators.

In the linear congruence method a sequence of random numbers is generated, using a starting value x_0, by

$$x_{i+1} = (a * x_i + b) \bmod(m), \tag{7.1}$$

with a, b appropriate integer constants and m the largest representable random number (integer), all operations are integer operations. The effect of $\bmod(m)$ is that digits "left of m" are irrelevant. The problem is, how to generate e.g. 1000 random numbers in parallel, i.e. in a pipeline, so that we get the same sequence as with scalar recursive computation. If we insert the relation for x_1 into x_2 we get from (7.1)

$$x_2 = (a_2 x_0 + b_2) \bmod(m) \tag{7.2}$$

with new "transfer coefficients" a_2, b_2 that relate x_2 to x_0 or x_{i+2} to x_i. With (7.2) and index shift we can express x_4 by x_2, introduce x_2 from (7.2) and get new coefficients a_4, b_4 that relate x_{i+4} to x_i. In the same way we can generate a_8, b_8; a_{16}, b_{16}; ..... until desired values a_n, b_n. Together with these coefficients we compute x_1 to x_n. Now we can generate the next n random numbers from the first n by

$$x_{n+1} = (a_n x_1 + b_n) \bmod(m)$$
$$\cdot$$
$$\cdot \qquad\qquad (7.3)$$
$$\cdot$$
$$x_{2n} = (a_n x_n + b_n) \bmod(m)$$

then the next set of n numbers and so on. The operations in (7.3) are independent and thus vectorizable with vector length n (and equally well parallelizable).

A faster method is the "exclusive or" method of Tausworthe [5]. The exclusive or is binary addition without carry and is denoted by (+). If we have generated a sequence of p random numbers, e.g. by the linear congruence method, we can generate a set of q new random numbers by

$$x_k = x_{k-p} \; (+) \; x_{k-q} \; , \qquad\qquad (7.4)$$

with q < p. Thus we combine two random numbers that are "back" by p and q for a new random number. This is executed as a vector operation with vector length q. Then the next set is generated and so on. An Internal Report [6] is in preparation where also corresponding references are given.

8. SUPERCOMPUTER SOFTWARE INDEPENDENT OF A SPECIAL ARCHITECTURE

The usual life-cycle of a supercomputer is 4 to 5 years, then it will be replaced by a new one, eventually with a different architecture. We had to replace our CYBER205 (memory-to-memory) by a Siemens/Fujitsu VP (register). But going from a CRAY Y-MP to a CRAY-2 or vice versa means also using different architectures. How can we design software that is as far as possible independent of a special architecture? In the preceding sections we have presented some rules that should guide us for the development of vector computer software. Software that is designed according to these rules is optimal for all types of vector computers, at least on the Fortran level. Here we repeat again these rules:

(1.3) Make vectors as long as possible (exception: priority of data locality for cache machines).

(3.4) Try to make your program 100% vectorizable with long and contiguous vectors.

(3.5) Separation of the selection and of the processing of the data.

(4.2) Select the optimal loop of a nest of loops for vectorization, adapt the data structure correspondingly.

(5.1) Sacrifice dummy operations to avoid individual treatment.

Examples of the application of these rules have been presented above in the corresponding sections.

A further rather general rule is the following

Rule: Develop a modular program structure. (8.1)

This is very important for vector computers. It will allow to replace easily modules that have been adapted to a special architecture, e.g. to a memory bottleneck or to a cache structure, by other modules for another architecture.

The weakness and limitations of Fortran do not allow a fully efficient use of vector computers. Therefore it may be necessary to design special machine-dependent or even assembler coded subroutines for compute-intensive kernels. Thus the basic program may be machine-independent, except these special subroutines. This leads to the following

Rule: Shift all machine-dependent parts of your program to a few special subroutines. (8.2)

Going to another architecture then means only exchanging these special subroutines. The FIDISOL program package for PDEs has been designed and adapted to different vector computers according to rules (8.1) and (8.2), see [2].

Rule (8.2) is also the philosophy of the international mathematical program libraries IMSL and NAG for the domain of linear algebra. Older linear algebra packages were based on the BLAS1 (basic linear algebra subprograms) that are vector vector operations. Later packages were based on BLAS2 [7] that are matrix-vector operations. But only BLAS3 [8] with matrix-matrix operations give full efficiency on vector computers because they allow an optimization over all nested loops. We wonder why the experts in this field needed such a long time to recognize this simple fact. The linear algebra packages of IMSL and NAG are completely based on the BLAS and thus up to this point machine-independent. It is the duty of the manufacturers of vector computers to provide optimal BLAS, eventually programmed in assembler code. This is exactly what is expressed by rule (8.2).

9. CONCLUDING REMARKS

The ultimate goal - and ambition - of the designer of a vector computer program must be to get as close as possible to the theoretical peak performance of the vector computer. If he is below, he must ask: Where are the lost cycles. Unfortunately he will meet in this investigation many hardware bottlenecks: All existing vector computers are more or less compromises between that what an excellent vector computer should be [9] and an old von Neumann type general purpose computer.

Concerning the software, here is a severe warning: Never believe the manufacturer! He will promise you the best compiler and vectorizer and tells you: It will make from your old program an excellent vector computer program. This can never be true because a compiler or vectorizer can only change code and never change data structures, remember the example of the Red-Black SOR.

Thus it is up to you to design optimal vector computer software. This will not be possible by adapting old software on the code-level. Only a redesign from scratch will result in optimal vector computer software, because the data structure is the key to high efficiency of the software. So the choice of the data structure is the starting point of a program design.

A final remark to multi-, makro-, mikro-, autotasking, i.e. to use several processors of a multiprocessor vector computer for the same job. This needs synchronization and certainly produces additional lost cycles. A justification is given only if a user needs all of a principal resource, e.g. the whole available memory. If he then used a single processor, the other processors would be idling and thus more cycles would be lost than if the program operates even with a bad multitasking. But then the user should have the possibility to reserve the whole multiprocessor for his job - and pay it. A mixture of multiprogramming and multitasking fails just for those jobs that need all of a principal resource. Thus the problem of multitasking is to minimize not only the lost cycles of a single processor but of all processors of a multiprocessor vector computer. It is clear that such a problem is still more architecture-dependent than a usual vector computer program. Presently we have to live with such poor architectures. How a user-friendly supercomputer should look like can be seen in reference [9].

10. REFERENCES

1. R.W. Hockney and C.R. Jesshope, *Parallel Computers 2*, Adam Hilger, Bristol 1988.
2. W. Schönauer, *Scientific Computing on Vector Computers*, North-Holland, Amsterdam 1987.
3. R. Weiss, H. Häfner and W. Schönauer, "Tuning the matrix-vector multiplication in diagonal form", in *Parallel Computing '89*, G.R. Joubert et al., eds., North-Holland, to appear.
4. H.A. van der Vorst and K. Dekker, "Vectorization of linear recurrence relations", *SIAM J. Sci. Stat. Comput.* 10:27-35 (1989).
5. R.C. Tausworthe, "Random numbers generated by linear recurrence modulo two", *Math. Comp.* 19:201-209 (1965).
6. W. Wälde, prospected title: "High Performance Random Number Generators on the Siemens VP Vector Computer Systems", to appear 1990 as Internal Report of the Computer Center of the University of Karlsruhe.
7. J.J. Dongarra, J.J. Du Croz, S.J. Hammarling and R.J. Hanson, "An extended set of Fortran Basic Linear Algebra Subprograms", *ACM Trans. Math. Software* 14:1-17 (1988).
8. J.J. Dongarra, J.J. Du Croz, I. Duff and S.J. Hammarling, "A set of level 3 Basic Linear Algebra Subprograms", Argonne National Laboratory, Mathematics and Computer Science Division, Technical Memorandum No. 88 (Revision 1) (1988).
9. W. Schönauer, "Could user-friendly supercomputers be designed?", in this book.

VECTORIZATION OF SOME GENERAL PURPOSE ALGORITHMS[1]

F. Brosens[*] and J.T. Devreese[**]

Department of Physics, University of Antwerp (UIA), Universiteits-plein 1, B-2610 Wilrijk-Antwerpen, Belgium

ABSTRACT

Some illustrative examples are given of the use of vectorization in practical circumstances, with working subroutines and programs, taken from existing scalar programs of general use. The need of vector extensions in the FORTRAN language is emphasized to realize global vectorization, which is beyond the capability of any auto-vectorizer.

The concepts of vectorization are illustrated for the user with a working knowledge of FORTRAN, and not for the specialist in a specific computational domain. Black boxes for vectorization applications are not provided: in practice, efficient vectorization will depend on the problem at hand, especially if one aims at **global** vectorization (throughout the program). Some eclectic problems which belong to the "common knowledge" are picked out, to illustrate how some rather simple modifications of the basic routines allow to realize vectorization through subprograms. The **algorithms with their conceptual description** are considered as adaptable building stones. In many cases, arguments of the scalar subprograms can be generalized to vectors, which allows efficient vectorization with minor modifications.

[1] Work supported by the Supercomputer Project of the NFWO (National Fund for Scientific Research, Belgium).
Partially performed in collaboration between the Management Unit of the Mathematical Model (MUMM) of the North Sea (Ministry of Public Health and Environment) and the "ALPHA"-Supercomputer-emulator project (NFWO-UIA).
[*] Senior Research Associate of the NFWO (National Fund for Scientific Research, Belgium).
[**] And: University of Antwerp (RUCA), Groenenborgerlaan 171, B-2020 Antwerpen (Belgium);
and: University of Technology, Eindhoven (The Netherlands).

I. INTRODUCTION

It looks quite natural that a vector computer provides the tools for handling vectors, not only at the level of the machine language, but also at the level of a user language like FORTRAN. ETA10 and CYBER-205 have provided specific extensions of the FORTRAN language (at the expense of portability of programs to other machines), whereas on other vector computers one has decided to keep the FORTRAN-77 standard (at the expense of flexibility in the vectorization). The latter option leaves the vectorization task to the computer (although the user can give directives), and the programmer should take care of writing his programs in such a way that the compiler can detect whether certain portions of the program are vectorizable.

The power of these vector extensions is merely discussed as tools for realizing global vectorization.[1] In fact, for efficient **global** vectorization, vector extensions at the FORTRAN level seem indispensable.

The idea of global vectorization has been applied to the problem of solving the time-dependent Hartree-Fock equation in the electron gas, in which a gain factor of 120 was obtained on a CYBER-205 by vectorization (see Ref. [1-3]).

It can not be denied that the extensive use of the vector extensions poses a serious transportability problem, which has become acute because of the commercial failure of ETA10. Since the programs as presented are developed for ETA10 and CYBER-205, they have to be modified for running on other machines. Nevertheless, it remains useful to develop vector programs with vectors as variables in mind, and to disentangle these vectors afterwards in the actual implementation of the program on a vector machine without vector extensions in the FORTRAN. It should also be noted that the ALPHA-project (see Ref. [3]) offers useful tools to realize the devectorization for existing vector programs.

Some illustrative examples are given of the use of vectorization in practical circumstances. The examples are taken from *"Numerical Recipes"* [4], because this book offers for each topic considered a certain amount of general discussion, of analytical mathematics, of discussion of algorithmics, and of actual implementations in the form of working computer routines.

As already mentioned, the actual implementations of the programs are written in the vector-FORTRAN of CYBER-205 and ETA10, hereafter denoted by FORTRAN-200. The standard scalar FORTRAN is referenced as FORTRAN-77.

II. THE VECTOR CONCEPT IN FORTRAN-200

A first prerequisite for vectorization is that the operations act on VECTORS. Generally speaking, a vector is a stream of values. On ETA10 and CYBER-205, a stream of values can only be realized with data (of the same type) in **contiguous** memory locations[2].

---

[1] Although the propagation of these vector extensions is temporarily slowed down due to the commercial failure of ETA 10, their power will again become effective with the FORTRAN-xx to come.

[2] These contiguous memory locations are indicated as "stride 1". For other vector computers than ETA10 and CYBER-205, also "stride n", with n a constant, is allowed, which refers to memory locations separated from each other by a constant amount of locations.

Since the compiler reserves contiguous memory locations (in well-defined order) to the elements of an ARRAY, a vector is sometimes confused with an array. The distinction between arrays and vectors is quite important for practical purposes.

Consider an array A with 100 elements, declared by the statement:

DIMENSION A(5,20)

In essence this declaration tells the compiler that the ARRAY A consists of 5 ' 20 elements, to be contiguously stored in memory in the order

| A(1,1) | A(2,1) | A(3,1) | A(4,1) | A(5,1) | A(1,2) | A(2,2) | ... | A(5,20) |

such that the element A(I,J) is located in the address $n+(I-1)+(J-1)*5$, where n denotes the address of the array element A(1,1). The array-notation in essence allows to identify specific elements out of a large number of variables. This is not a vector, but a vector can be constructed with a set of elements in contiguous memory locations. E.g.

| A(3,1) | A(4,1) | A(5,1) | A(1,2) |

can be considered as a vector of length 4 if this stream is input or output for a sequence of operations.

It is clear that FORTRAN-77 is not adapted to handle vectors: it can only refer to (addresses of) individual variables.[3] The vector-extensions of FORTRAN-200 are a very logical implementation, which form a kind of a (software) interface between the vector architecture and the programmers's needs to explicitly address the vector processing facility.

The disadvantage of this approach is of course the fact that programs in FORTRAN-200 are not transportable to many other vector machines; other vendors therefore have decided to keep with the standard FORTRAN, and to await the new FORTRAN-XX standard to come, which will incorporate a vector concept in the language.

The essential difference between FORTRAN-200 and FORTRAN-77 is the fact that FORTRAN-200 has language elements to handle vectors as a variable , where a vector in FORTRAN-200 is a sequence of values that are stored in contiguous memory locations.

Furthermore, some tools are provided to handle this new type of a variable efficiently, e.g.
- DESCRIPTOR specification statement;
- ASSIGN descriptor and FREE statement;
- vector assignment statement;
- WHERE statement and WHERE ... OTHERWISE ... ENDWHERE block;
- vector FUNCTION definition.

---

[3] Even if arrays are passed as arguments for subprograms, only the address of the first array element is passed. In older programs this fact is often used to bypass the need of giving the actual dimensions of an adjustable array explicitly; in such subprograms one often encounters array declarations of the form DIMENSION X(1).

Some scalar extensions are also provided, but there main use is related to the vector handling capacity, e.g.
- ROWWISE specification statement;
- BIT specification statement;

FORTRAN-200 (like the FORTRAN on most machines) also provides some other (not specifically vector-related) extensions (e.g. HALF PRECISION data type; HEX-ADECIMAL constants, HOLLERITH constants with left- or right-justification, ARRAY ASSIGNMENT STATEMENTS, SUBARRAY REFERENCES, ...) which are not related to specific vector applications, and which are not considered or used here at all.

III. MAIN VECTOR EXTENSIONS IN FORTRAN-200

The vector extensions of FORTRAN-200 provide rather important vectorization tools, especially for realizing global vectorization. The auto-vectorizers (and more sophisticated vectorizers like KAP or VAST) also are important tools, but at a different level: the local level of the DO-loops in existing scalar code. A good understanding of the vector concept helps to understand the role of the vectorizers: they do nothing that can not be done by the users; on the contrary, they only vectorize to a much more limited extent, because it is much more difficult to discover the vectorization possibilities from the actual scalar code than from the program concept.

III.A. Vector Variables and Vector Assignments

As already indicated in the section on the concept of a vector, three elements are required to characterize a vector:
- the type of its elements;
- the address of its first element;
- its length[4].

The type of the variables is implicitly or explicitly known from the type declaration statements[5]. Essentially, two methods are provided in FORTRAN-200 to give the information required: the explicit and the implicit vector reference.

III.A.1. Explicit vector reference

The simplest language element in FORTRAN-200 for indicating a vector applies to arrays: one gives the array element (i.e. array name and the appropriate indices) which is the first element of the vector, and the number of elements in the vector as follows:

```
array-name({list of indices}; vector-length)
```

---

[4] On ETA10 and CYBER-205, the maximum vector length is 65535 for 64-bit variables, whereas it is limited to 32767 for 128-bit variables. The maximum vector length should never be exceeded, nor should one implicitly try to impose a non-positive vector length: the results would be **unpredictable**!!

[5] Complex or double precision vectors are often exceptions. Before using them, one should check in the manual whether the intended manipulation indeed applies.

The semicolon $\boxed{;}$ is very characteristic for vector notations: it is followed by a reference to the vector length. A similar construction will be used for vector functions (see below):

```
function-name({list of arguments}; vector-length)
```

Consider e.g.

```
DIMENSION A(10)
DIMENSION C(2,3)
```

$A(2;9)$ then means a vector with the following 9 elements: $A(2)$, $A(3)$, ..., $A(10)$. $C(1,2;4)$ means a vector with 4 elements: $C(1,2)$, $C(2,2)$, $C(1,3)$, $C(2,3)$. **But $C(1,2;5)$ would lead to unpredictable results!!!**

Vector references are building stones for vector expressions and vector statements (similar as for scalar variables, although their internal treatment in a vector computer differs substantially). E.g. the vector arithmetic expression $A(2;9)**2$ returns a vector with the elements $A(2)**2$, $A(3)**2$, ..., $A(10)**2$.

Vector arithmetic expressions can be used for constructing compound vector expressions, similarly as with scalar variables. An experienced FORTRAN user will have little difficulty in correctly interpreting a combined vector expression like:
$((A(2;9)+2)**(1./3.))/(A(2;9)-3)+A(2;9))$.

In all these constructions, vector functions can be used in the same way as the explicit vectors builded from array elements.

Vector arithmetic expressions can be used in arithmetic vector assignment statements, which are of the form:

```
vector = vector-expression
          or
vector = scalar expression
```

If the right-hand side is a vector expression, then the first element of the output vector will be assigned the value of the first element in the right-hand side vector, and so on. If the right-hand side is a scalar expression, then each element of the output vector is assigned the value of the scalar expression.

E.g., the following program:

```
DIMENSION I(10)/1,2,3,4,5,6,7,8,9,10/
I(1;10)=I(1;10)**2
I(2;4)=-I(2;4)
I(8;3)=1
...
```

will contain the following elements of I after these manipulations:
$I(1)=1$, $I(2)=-4$, $I(3)=-9$, $I(4)=-16$, $I(5)=-25$, $I(6)=36$, $I(7)=49$, $I(8)=1$, $I(9)=1$, $I(10)=1$.

As an example of the use of the vector notation, consider the matrix multiplication $C = A \times B$, where A is a $L \times M$ matrix, and B a $M \times N$ matrix.

```
            DIMENSION A(L,M),B(M,N),C(L,N)
C **        Initialization ....
C           Scalar matrix multiplication
            DO 1 LL=1,L
              DO 1 NN=1,N
                C(LL,NN)=0
                DO 1 MM=1,M
1                  C(LL,NN)=C(LL,NN)+A(LL,MM)*B(MM,NN)
C           Vector matrix multiplication
            C(1,1;L*N)=0
            DO 21 NN=1,N
              DO 1 MM=1,M
21               C(1,NN;L)=C(1,NN;L)+A(1,MM;L)*B(MM,NN)
```

Matrix multiplication

		Execution time (in sec.) for matrix multiplication	
	$C = A * B$	$A: L \times M;\ B: M \times N$	
$L =$	2	2	100
$M =$	3	100	100
$N =$	2	1	100
Scalar	0.000046	0.000573	2.62520
Vector	0.000024	0.000252	0.02996

$$C_{l,n} = \sum_{m=1}^{M} A_{l,m} B_{m,n}$$

In the program above, both a scalar and a vector version are implemented. The table with the comparison of the timing results is also given, as obtained on the ETA10-P of the CAMME-project of the Ministry of Public Health and Environment.

Similarly as for arithmetic operations, a bit vector is obtained if the relational operators $\boxed{\text{.GT., .NE., ...}}$ are used between two numeric vectors, or between a numeric vector and a numeric scalar (analogous to the logical expressions for scalar variables). Here again, if two vectors are compared, the comparison is made element by element. If a vector is compared to a scalar, each element of the vector is compared to the scalar. The conversion rules are the same as for scalar variables. These can be used in bit vector expressions and bit vector assignment statements.

III.A.2. Implicit vector reference

The explicit vector reference, as discussed above, requires that an array or vector function is explicitly defined in the program. For the vectors, defined on an array, one

28

might feel inclined to consider the vector assignment statements as a short hand notation for DO-loops.

The full power of vectorization in FORTRAN-200 becomes available with the implicit vector reference, which introduces a different kind of variable, to be declared by the DESCRIPTOR statement in the declarative part of the program:

```
DESCRIPTOR var1,...,varN
```

where var1, ..., varN are BIT, INTEGER, HALFPRECISION, REAL or COMPLEX variables or even array declarators, which will denote vectors.

As already mentioned above, a vector is specified by the type of its elements (BIT, INTEGER, ...), the address of the first element, and the length of the vector. The type of the elements of a vector is defined by the type (implicit or explicit) of the vector variable.

E.g. the declaration statements:

```
REAL I
DIMENSION A(5)
DESCRIPTOR I,A
```

declare that I will denote a vector with real elements, and that *each* of the array elements A(1), ..., A(5) will be a vector with real elements (since by default A is a real variable in FORTRAN).

The remaining requirements to define the vector completely are associated to the vector variables by the ASSIGN statement, for which two forms are available:

```
ASSIGN var1, explicit vector
     or
ASSIGN var1, .DYN. integer expression
```

In the first form, where an explicit vector is assigned to a descriptor var1, the descriptor can be used as a short hand notation for this vector reference. The first element of the explicit vector is then also the first element of the implicit vector, and their lenghts are equal, which means that both vectors are equivalent (until the descriptor is possibly redefined by another ASSIGN statement in the same subprogram).

With the dynamical assignment, only the length of the vector is given explicitly. The computer determines the location of the first element, depending on the current occupation of memory space. This form is particularly useful for introducing auxiliary vectors, and for defining vector functions (see below). This dynamical space can be released by the FREE statement, which is of the form:

```
FREE
PROGRAM VSQUARE
DIMENSION A(10,10),B(10)
```

```
      DESCRIPTOR B
         DO 1 J=1,10
            ASSIGN B(J),A(1,J;10)
               B(J)=J
1
      DO 2 J=2,10,2
2     CALL SQUARE(B(J),B(J-1))
      END

      SUBROUTINE SQUARE(A,B)
      DESCRIPTOR A,B
      B=A**2
      END
```

Example program $\boxed{\text{VSQUARE}}$

Also upon RETURNing from a subprogram, the dynamical space of that subprogram is automatically released.

Note that it is impossible to directly access individual elements of a dynamically assigned vector, since no variables are available to refer to these elements!!

Note also that vectors (in any valid form) can be used as **arguments** in functions or subroutines. E.g. consider the (artificial) program VSQUARE.

The subroutine SQUARE(A,B) in this example returns in the output vector B the elements squared of the input vector A. (Note that no ASSIGN statement is required in this subroutine, because the vectors are supposed to be defined in the calling program.)

Because of the statements ASSIGN B(J),A(1,J;10) for J=1, ...,10, the vector B(J) in the main routine denotes the Jth column of the 10×10 matrix A. In the first DO-loop, each element in this column is given the value J (automatically converted to type real). Thus after the first DO-loop, the array A can be represented as:

$$A = \begin{pmatrix} 1. & 2. & 3. & 4. & 5. & 6. & 7. & 8. & 9. & 10. \\ 1. & 2. & 3. & 4. & 5. & 6. & 7. & 8. & 9. & 10. \\ \vdots & & & & & & & & & \\ 1. & 2. & 3. & 4. & 5. & 6. & 7. & 8. & 9. & 10. \end{pmatrix}$$

In the second DO-loop, the even columns are transmitted as input argument to the routine SQUARE, which returns as output the square of these elements in the preceding column. Thus after this second loop, the array A is converted into:

$$A = \begin{pmatrix} 4. & 2. & 16. & 4. & 36. & 6. & 64. & 8. & 100. & 10. \\ 4. & 2. & 16. & 4. & 36. & 6. & 64. & 8. & 100. & 10. \\ \vdots & & & & & & & & & \\ 4. & 2. & 16. & 4. & 36. & 6. & 64. & 8. & 100. & 10. \end{pmatrix}$$

Less artificial examples will be given below.

III.A.3. Vector functions

In the calling program, a vector function can be referenced as an explicit vector, except that a number of arguments can be passed. E.g. suppose that a vector function `VSIN(X;length)` exists, with returns as a vector the sine function of all the elements of a vector X. Then

        VSIN(A(1;N);N)

is a vector expression in the calling program, with the first element given by `SIN(A(1))`, the second by `SIN(A(2))`, etc.

In the calling program, a second possibility exists for referring to a vector function, nl. to pass a vector of the appropriate length as the output argument, instead of the length only, e.g.:

        VSIN(A(1;N);B(1;N))

In this case, `B(1;N)` will be used as a vector to store the values of the vector function. This form is of course interesting for handling implicit vector references with variable length, sometimes even unknown to the programmer. But it has also an interesting "side-effect": the values of the vector are stored in the output vector for future use. E.g.
`B(1;N)=(VSIN(A(1;N);N)**2` is equivalent to
`B(1;N)=(VSIN(A(1;N);B(1;N))**2` but `B(1;N)=(VSIN(A(1;N);C(1;N))**2`
returns in `C(1;N)` the sine of the elements of `A(1;N)`, and in `B(1;N)` the square of the elements of `C(1;N)`.

To summarize, the **calling** sequence of a vector function is of the form:

        Vector-function(arg1,...argN;output-vector)
    or
        Vector-function(arg1,...argN;length)

To define a vector function, the user can essentially proceed like with the definition of other functions, except for the fact that the closing parenthesis) in the FUNCTION statement has to be replaced by ;*). This indicates that the output vector is of adjustable length (and will have to be defined in the calling program). **Furthermore, the function name has to be declared a descriptor.** The essential elements to define a vector function are thus:

        FUNCTION name(arg1,...argN;*)
        DESCRIPTOR name
        ....
        END

Consider e.g. a polynomial $P(x)$ of degree N-1, defined by a stored array of coefficients $C(J)$, $J = 1,...,N$, where $C(J)$ is the coefficient of x^{J-1}. If one is interested in the value of the polynomial for a large number of arguments x, we give a rather simple vector function `VPOL(C,N,X;*)`, to be compared to the corresponding scalar version `POL(C,N,X)`.

```
      FUNCTION VPOL(C,N,X;*)
      DESCRIPTOR X,VPOL
      DIMENSION C(N)
C RETURNS AS VECTOR VPOL THE VALUES OF C(1)+...C(N)*X**(N-1)
C FOR ALL ELEMENTS OF THE VECTOR X
      VPOL=C(N)
      DO 11 J=N-1,1,-1
11    VPOL=VPOL*X+C(J)
      RETURN
      END
```

Vector polynomial function $\boxed{\text{VPOL}}$

```
      FUNCTION POL(C,N,X)
      DIMENSION C(N)
C RETURNS AS POL THE VALUES OF C(1)+...C(N)*X**(N-1)
      POL=C(N)
      DO 11 J=N-1,1,-1
11    POL=POL*X+C(J)
      RETURN
      END
```

Scalar polynomial function $\boxed{\text{POL}}$

Execution time (in sec.) for 1001
values of x
of $5x^2-6x+1$

With POL 0.01464	With VPOL 0.00007

To evaluate e.g. $5x^2-6x+1$ in 1001 equidistant points x, with $x_1 = 0$,..., $x_{1001} = 1$, we used both the scalar and the vector version, and compared the CPU-time needed on the ETA10-P of the CAMME-project of the Ministry of Public Health and Environment.

One thus immediately realizes that the vectorization involves much more than a compact notation: the vector version is organized in a stream of data which is very efficiently handled by the vector processor.

It should also be emphasized that neither the autovectorizer, nor KAP or VAST are able to vectorize the scalar version of this program properly. These "vectorizers" only try to treat the DO-loops **inside** a module, i.e. the DO-loop in the routine POL on the coefficients C(J), which is recursive in nature. The global vectorization, as realized in VPOL, can not be realized by any existing vectorizer.

III.B. Vector Flow Control

The vectorization tools considered until now, only handled manipulations on vectors without branching. Nevertheless, branching is one of the major programming tools.

In a sense, branching is almost contradictory to the concept of vectorization, since it breaks the contiguous stream of data, which is the essence of vector handling. The vector flow control facilities however only act on the output of a vector manipulation: certain elements of the output are not allowed to be stored in memory.

FORTRAN-200 provides two (strongly related) means of controlling the output of a vector operation: the `WHERE` statement and the `WHERE...ENDWHERE` block (similar to the `IF` statement and the `IF...ENDIF` block in `FORTRAN-77`).

The `WHERE` statement is of the form:

```
      WHERE (bit vector expression) vector assignment
e.g.
      WHERE (X(1;10).NE.0) X(1;10)=1./X(1;10)
```

The bit vector expression and the vector assignment statement should have result vectors of the same length. In contrast to the `IF` statement, the vector operations in a `WHERE` statement are **executed** for all the vector elements, but the result is not **stored** in the result vector for those elements which do not have a corresponding element `'1'` in the bit vector expression.

The `WHERE...ENDWHERE` block is of the form

```
      WHERE (bit vector expression)
        vector assignment statement(s)
      ENDWHERE
```

or

```
      WHERE (bit vector expression)
        vector assignment statement(s)
      OTHERWISE
        vector assignment statement(s)
      ENDWHERE
```

The meaning of these blocks is quite similar as with `IF...ENDIF` blocks. The results inside the `WHERE` part are only stored in memory if the corresponding elements in the bit vector expression are `'1'`, whereas the `'0'` values result in storing the corresponding elements of the assignments in the `OTHERWISE` part of the block (if present).

WARNING: Only very simple vector statements are allowed inside a `WHERE` statement or block, nl. the elementary operations addition, substraction, mutiplication and division (**no exponentiation!**) and the vector functions references `VFLOAT`, `VIFIX`, `VINT`, `VAINT`, `VSQRT`, `VABS` and `VIABS`.

IV. INTRINSIC FUNCTIONS

As a consequence of the extensions of FORTRAN-200 as compared to FORTRAN-77, a large set of intrinsic functions is provided. For a full discussion in detail the manual should be consulted. We only handle some typical examples, which are very useful for realizing global vectorization of a program.

IV.A. Scalar Functions with Scalar Arguments

Because of the introduction of HALF PRECISION variables, several extra functions are provided, most of which have the prefix H. E.g. HSIN(X) returns the half precision value of the sine function applied to the half precision value X. However, most of these functions also have a generic version (e.g. SIN for the sine) and in most circumstances no specific half-precision function is explicitly required. (The situation is very much similar to the one for DOUBLE PRECISION variables.)

The introduction of the BIT variables also leads to the conversion function BTOL and LTOB. The function BTOL has a bit-variable as an argument, and returns the corresponding logical value .TRUE. or .FALSE.. Similarly, LTOB requires an argument of type logical, and returns the corresponding bit-value.

IV.B. The Intrinsic V-functions

These functions are the natural vector extensions of the standard intrinsic functions for arithmetic operations and type-conversions. For most of these intrinsic functions with calling sequence of the form 'function-name(argument)', the standard rule for calling the vector extension is:

- the function name is preceded by the prefix V, e.g. VSIN, VCOS, VABS, VEXP, VLOG, ...;
- the argument is replaced by a vector argument;
- the closing parenthesis ')' is replaced by ';output)', where output is an output vector **or** the length of the output vector.

E.g. the statement
B(1;10)=VSIN(A(1;10);10)
returns
B(1)=SIN(A(1)), ..., B(10)=SIN(A(10)).

As for the scalar corresponding functions, in most cases there exists a generic vector function (like VSIN), which accepts arguments of several types, as well as some specific versions (like e.g. VHSIN, which requires a half precision vector argument).

IV.C. The Intrinsic Q8-functions

The intrinsic Q8-functions result in highly optimized machine code for several non-trivial purposes, e.g. to handle recursive vector manipulations, to extract information on the length of a vector or on specific elements, to reorganize vectors, etc.

```
FUNCTION FACTLN(N)
IF (N.LT.0) STOP ' FACTLN: N<0'
FACTLN=0.
IF (N.GT.1) THEN
FACTLN=Q8SSUM(VLOG(Q8VINTL(2.,1.;N-1);N-1))
ENDIF
RETURN
END
```

Program $\boxed{\text{FACTLN}}$ for log($N!$)

As a general rule, the functions with prefix $\boxed{\text{Q8S}}$ return a scalar result, whereas the prefix $\boxed{\text{Q8V}}$ indicates that the result is a vector. Again, in all the Q8V-functions, the length of the output vector can be given as output argument (after the ; delimiter), as well as a vector reference (of the correct type). In the following descriptions, we will use the short hand notation ;N for the output argument in both cases.

Also, we will not give a full overview, but limit the discussion to some typical examples. For a complete overview and a description in full detail, we refer to the manual.

IV.C.1. Initialization of a vector

- Q8VINTL(X1,DX;N) (generic) performs the initialization of a vector of length N, starting with X1 in the first element, with an increment DX for the following elements. (Cfr. the scalar loop X(J) = X1 + (J-1)*DX.)

- Q8VMKO(N1,N2;N) results in a periodically filled bitvector of length N with a pattern of N1 bit values '1' followed by N2-N1 bit values '0'.

- Q8VMKZ(N1,N2;N) is analogous to Q8VMKO, but with the bit values '1' and '0' interchanged.

IV.C.2. Extracting scalar information from vectors

- Q8SCNT(BIT) returns the number of '1' values in the input vector BIT.

- Q8SLEN(V) returns the length of a vector V of any type.

- Q8SDOT(U,V) (generic) returns the "dot-product" $\sum_{i=1}^{N} U_i V_i$ of two vectors of equal length N. This function is useful for many numerical quadrature rules $\int_a^b f(x)\,dx \approx \sum_{i=1}^{N} w_i f(x_i)$.

- Q8SSUM(V) (generic) returns the sum of all the elements of the vector V. The program FACTLN to calculate log($N!$) = $\sum_{i=1}^{N} \log(J)$ illustrates its use.

- Q8SPROD(V) (generic) returns the product of all elements of the vector V, as illustrated by the programs FACTRL and FACT2L to calculate $N!$ and (2N-1)!! resp. as a floating point number.

```
FUNCTION FACTRL(N)
IF (N.LT.0) STOP 'FACTORIAL OF NEGATIVE INTEGER'
FACTRL=1.
IF (N.GT.1) FACTRL=Q8SPROD((Q8VINTL(2.,1.;N-1))
RETURN
END
```

Program FACTRL to calculate $N!$ as a floating point number.

```
FUNCTION FACT2L(N)
IF (N.LT.0) STOP 'FACTORIAL OF NEGATIVE INTEGER'
FACT2L=1.
IF (N.GT.1) FACT2L=Q8SPROD((Q8VINTL(3.,2.;N-1))
RETURN
END
```

Program FACT2L to calculate $(2N-1)!!$ as a floating point number.

IV.C.3. Extracting vector information from vectors

We only mention a few examples. E.g.

- Q8VADJM(V;N) (generic) returns as i^{th} element the value of $(V_i+V_{i+1})/2$.

- Q8VDELT(V;N) (generic) returns as i^{th} element the value of $(V_{i+1}-V_i)$.

- Q8VAVG(U,V;N) (generic) returns as i^{th} element the value of $(U_i+V_i)/2$.

IV.C.4. Reversion, compression, expansion, merging, ... of vectors

- Q8VREV(V;N) produces an output vector with the elements of the input vector V in reverse order.

- Q8VDCMPR(V1,V2,BIT;N) (generic) the result vector consists of the elements of V2, **except** for the elements where BIT has corresponding 1-bits. For these elements, consecutive elements of V1 are filled out in the result vector.

As an example, consider the situation where some operations are to be performed on the Jth row of an array A(N,M). We explicitly give a program part that can be used to extract and re-insert this row into the array.

- Q8VCMPRS(V,BIT;N) requires an input vector BIT of type bit, and an input vector V of type integer, real or half precision. Both input vectors must have the same length. The output vector consists of the elements of V with corresponding 1 bits in the bit vector.

This function is very useful if a WHERE statement (or block) would require too many useless operations, or if functions are involved which are not allowed in a WHERE statement.

```
        DIMENSION A(N,M)
        BIT BIT
        DESCRIPTOR ROWJ,BIT
        ...
C Extract row J
        JAUX=(M-1)*N+1
        ASSIGN BIT,.DYN.JAUX
        ASSIGN ROWJ,.DYN.M
        BIT=Q8VMK0(1,N;BIT)
        ROWJ=Q8VCMPRS(A(J,1;JAUX),BIT;M)
C Perform operations on ROWJ
C and put ROWJ back into array A
        A(J,1;JAUX)=Q8VDCMPR(ROWJ,A(J,1;JAUX),BIT;JAUX)
        ...
```

Extract and reinsert a row in an array.

```
        DIMENSION A(L,M),B(M,N),C(L,N)
        DESCRIPTOR AUX
        ASSIGN AUX,.DYN.M
C **       Initialization of A and B
        KAUX=(M-1)*L+1
        DO 11 LL=1,L
          AUX=Q8VCMPRS(A(LL,1;KAUX),Q8VMK0(1,L;KAUX);AUX)
          DO 11 NN=1,N
11          C(LL,NN)=Q8SDOT(AUX,B(1,NN;M))
```

Matrix multiplication $C = A \times B$ using Q8VCMPRS.

Compression can e.g. be used to develop an alternative method for the matrix multiplication $C = A \times B$, where A is a $L \times M$ matrix, and B a $M \times N$ matrix.

$$C_{l,n} = \sum_{m=1}^{M} A_{l,m} B_{m,n}$$

Each element of C is a dot product of a row from A and a column from B. By using the periodic bit-function Q8VMK0 with the periodicity L as the control vector in Q8VCMPRS, the rows of A can be extracted in an auxiliary vector, as explicitly illustrated in an implemented program part.

Before, we discussed a method for matrix multiplication with respect to the columns in C, which we here denote as "direct". Compression is only favourable if the rows in A are substantially longer than the columns, and if the columns in B are much larger than the rows, as is illustrated in the timing table for different dimensions.

IV.C.5. Gather and scatter operations

These operations deserve special attention, because of their efficient capacity of rearranging vectors in almost arbitrary order. In the following functions, V denotes an input vector of type integer, half precision or real, of the same type as the output vector.

	Execution time (in sec.) for matrix multiplication		
	$C = A * B$	$A: L \times M$; $B: M \times N$	
$L =$	2	2	100
$M =$	3	100	100
$N =$	2	1	100
Scalar	0.000046	0.000573	2.62520
Vector	0.000024	0.000252	0.02996
Q8VCMPRS	0.000036	0.000046	0.08332

- Q8VGATHR(V,I;N) returns an output vector, of which the Kth element is given by V(I(K)), where the index vector I is an input vector of type integer, an its length is equal to the length of the output vector.

- Q8VSCATR(V,I;N) essentially performs the inverse operation as compared to the gather function. If U denotes the output vector, the result can be translated as U(I(K)) = V(K), with K running from 1 to the number of elements in the integer input vector I.

- Q8VGATHP(V,I,M;N) assigns the values of the elements V(1), V(1+I), V(1+2*I) ... V(1+(M-1)*I) to the first M elements of the output vector. The integer I thus denotes the periodicity, and M the number of elements which are inserted in the output vector.

The statement VOUT = Q8VGATHP(V,I,M;N) with VOUT and V one-dimensional arrays, thus corresponds to the following DO-loop:

```
      DO 1 J=1,M
1     VOUT(J)=V(1+(J-1)*I)
```

- Q8VSCATP(V,I,M;N) is analogous with the periodic gathering, with the difference that the periodicity I is now imposed on the elements of the output vector. (In this particular case, the input vector V can be replaced by a scalar, of the same type as the output vector.)

The statement VOUT = Q8VSCATP(V,I,M;N) } with VOUT and V one-dimensional arrays, corresponds to the following DO-loop:

```
      DO 1 J=1,M
1     VOUT(1+(J-1)*I)=V(J)
```

```
      DIMENSION A(M,N),AT(N,M)
      DO 1 J=1,M
1     AT(1,J;N)=Q8VGATHR(A(1,1;M*N),Q8VINTL(J,M;N);N)
```

Transposing a matrix A into AT using Q8VGATHR.


```
      DIMENSION A(M,N),AT(N,M)
      DO 1 J=1,M
1     AT(1,J;N)=Q8VGATHP(A(J,1;(N-1)*M+1),M,N;N)
```

Transposing a matrix A into AT using Q8VGATHP.


```
      DIMENSION A(M,N),AT(N,M)
      DO 1 J=1,N
      AT(J,1;(M-1)*N+1)=
=        Q8VSCATP(A(1,J;M),N,M;AT(J,1;(M-1)*N+1))
1     CONTINUE
```

Transposing a matrix A into AT using Q8VSCATP.

Programmers who are familiar with solving sets of linear equations will immediately realize the usefulness of these functions: they allow the rearranging of a matrix (and afterwards the restauration) under the control of an index vector governed by the pivoting.

As an elementary illustration, we give below the construction of the transpose array AT(N,M) of an array A(M,N) with the help of the gather function: row by row, an index vector is constructed which points to the element of the rows in A, and then is used for the gather operation. (One might also use the scatter function, which will be more efficient if the columns are shorter than the rows.)

However, in many situations one has to collect or distribute elements periodically, rather than with a random index vector. The example for transposing a matrix is in fact also typical for this periodicity. The Jth column of the transposed matrix AT has N elements, beginning with the element AT(1,J) = A(J,1), and the periodicity in the gather function equals the number M of elements in a column of A.

Periodic scattering can also be used for transposing a matrix. The Jth column of A with M elements is scattered with periodicity N into AT, beginning with the element AT(J,1) = A(1,J).

In the previous examples of gather and scatter operations, explicit matrices were associated with the vectors. But if only a vector without matrix-representation is to be handled, periodicity is not always helpful. Consider e.g. the rather simple problem of a vector XM with M elements, which is to be expanded into a vector XNM of length N*M as follows:

$$\overbrace{\hspace{2cm}}^{N} \qquad \overbrace{\hspace{2cm}}^{N} \qquad \qquad \overbrace{\hspace{2cm}}^{N}$$

XNM: $\quad x_1 \quad \ldots \quad x_1 \quad x_2 \quad \ldots \quad x_2 \quad \ldots \quad x_M \quad \ldots \quad x_M$

```
      FUNCTION VROWS(X,N;*)
C With input vector X_1,X_2, ..., X_M construct a vector VROWS
C as N elements X_1, N  elements X_2, ... , N elements X_M
      DESCRIPTOR VROWS,X,I
      M=Q8SLEN(X)
      NM=Q8SLEN(VROWS)
      IF (NM.NE.M*N.OR.NM.EQ.0)STOP 'INCORRECT LENGTH IN VROWS'
      ASSIGN I,.DYN.NM
      I=1+Q8VINTL(0,1;NM)/N
      VROWS=Q8VGATHR(X,I;NM)
      RETURN
      END
```

Vector function $\boxed{\text{VROWS}}$

In matrix notation, this means that we would like to construct an array XNM(N,M) from a row XM(M) as follows:

$$
XM = (x_1 \ \ x_2 \ \ ... \ \ x_M) \Rightarrow XNM =
\begin{array}{c} 1 \\ 2 \\ \cdot \\ \cdot \\ N \end{array}
\left(
\begin{array}{cccc}
x_1 & x_2 & ... & x_M \\
x_1 & x_2 & ... & x_M \\
\cdot & & & \\
\cdot & & & \\
x_1 & x_2 & ... & x_M
\end{array}
\right)
$$

The vector function VROWS performs this operation using Q8VGATHR, with an index vector constructed in the appropriate way.

A similar program VCOLS is also given, which can be used for copying a vector N times, which in matrix notation amounts to the following construction of an array XMN(M,N):

$$
XM =
\left(
\begin{array}{c}
x_1 \\ x_2 \\ \cdot \\ \cdot \\ \cdot \\ x_M
\end{array}
\right)
\Rightarrow XNM =
\begin{array}{cccc} 1 & 2 & & N \end{array}
\left(
\begin{array}{cccc}
x_1 & x_1 & ... & x_1 \\
x_2 & x_2 & ... & x_2 \\
\cdot & & & \\
\cdot & & & \\
x_M & x_M & ... & x_M
\end{array}
\right)
$$

As a typical example where this kind of vector prolongation is useful, we mention the vectorization of two-dimensional (and multi-dimensional) functions F(X,Y), needed on a grid of points X(I), I=1,2, ..., M and Y(J), J=1,2, ..., M. If the function is vectorizable, vectors of length M*N can be constructed by using VROWS for the X-values and VCOLS for the Y-values (or vice versa). For multi-dimensional functions, this process can be repeated several times.

V. PRACTICAL EXAMPLES

Some illustrative examples of the use of vectorization in practical circumstances are given, for some scalar routines described in Ref. [4], because this book offers for each topic considered a certain amount of general discussion, of analytical mathematics, of discussion of algorithmics, and of actual implementations in the form of working computer routines.

```
      FUNCTION VCOLS(X,N;*)
C With input vector X_1,X_2, ..., X_M construct a vector VCOLS
C with N of these vectors consecutively
      DESCRIPTOR VCOLS,X,I,IM
      M=Q8SLEN(X)
      NM=Q8SLEN(VCOLS)
      IF (NM.NE.M*N.OR.NM.EQ.0)STOP 'INCORRECT LENGTH IN VCOLS'
      ASSIGN I,.DYN.NM
      ASSIGN IM,.DYN.NM
      I=Q8VINTL(0,1;NM)
      IM=M
      I=1+VMOD(I,IM;NM)
      VCOLS=Q8VGATHR(X,I;NM)
      RETURN
      END
```

Vector function $\boxed{\text{VCOLS}}$

To illustrate the concepts of vectorization for the user (not for the specialist in a specific computational domain), we picked out some problems belonging to the "common knowledge", and illustrate how some rather simple modifications of the basic routines allow to realize vectorization through subprograms. One simply has to consider the **algorithms with their conceptual description** as adaptable building stones. In many cases, it is not too difficult to generalize the arguments of the subprograms to descriptors, and to apply the same techniques as in scalar mode on vectors.

All comparisons for the CPU time in scalar and vector operation in the following examples were performed on the ETA10-P of the CAMME-project of the Ministry of Public Health and Environment.

V.A. Integration with Equally-Spaced Abcissas

A typical example of a "work horse" for general-purpose use is the extended trapezoidal rule:

$$\int_{x_1=a}^{x_M=b} f(x)dx = h[\frac{1}{2} f_1 + f_2 + ... + f_{M-1} + \frac{1}{2} f_M] + \theta \left(\frac{(b-a)^3 f''}{M^2} \right)$$

(with $h = \frac{b-a}{M-1}$ the length of the intervals).

The procedure for implementing this rule for a fixed function f, as proposed in Ref. [4], is to start with both endpoints a and b alone, and to double the number of intervals in each stage of refinement (giving $M = 2^{(N-1)}$ intermediate points x_i in the Nth stage of refinement). This procedure benefits at any stage from the previous function evaluations. The routine TRAPZD computes the Nth stage of refinement for a function FUNC, with the result given in S.

```
      SUBROUTINE TRAPZD(FUNC,A,B,S,N)
      EXTERNAL FUNC
      SAVE IT
      IF (N.EQ.1) THEN
        S=0.5*(B-A)*(FUNC(A)+FUNC(B))
        IT=1
      ELSE
        TNM=IT
        DEL=(B-A)/TNM
        X=A+0.5*DEL
        SUM=0.
        DO 11 J=1,IT
          SUM=SUM+FUNC(X)
          X=X+DEL
11        CONTINUE
        S=0.5*(S+(B-A)*SUM/TNM)
        IT=2*IT
      ENDIF
      RETURN
      END
```

Program ⎢TRAPZD⎥ from Ref. [4].

If the function FUNC can be defined as a vector function, TRAPZD can be vectorized with very good results. Note however that in the first 4 stages of refinement, the vectorization would be a waste of time, since these stages require 2, 1, 2 and 4 function evaluations resp. It is then more efficient to calculate these 9 function values at once, store the results in an array, and pick out the required function values for the stage of refinement at hand. The implementation is given in the routine VTRAPZD.

It should be emphasized that the scalar part in the program VTRAPZD serves the only purpose to keep the routine compatible with all driver routines which previously used TRAPZD, with the same meaning for all the arguments. The cases where the trapezoidal rule gives sufficient accuracy with less then 9 points are rather exceptional.

For testing purposes, we integrated $x^2(x^2-2)\sin(x)$ from 0 to $\pi/2$. The vectorization of this integrand is very simple.

Although the modifications are minor, the vectorization turns out to be very successful as is manifest from the corresponding timing table.

One might (correctly) object that this program on its own is not an appropriate integration program, since it is based on the number of iterations, not on the accuracy of the result. This test however was only done to show how appreciable speed up factors can be obtained with simple means. For reference, we also give the timing results with two integration routines aiming at a maximum fractional error of 10^{-6}, based on the successive stages of refinement considered above. These two routines, nl. QTRAP and QSIMP are given below, and refer resp. to the extended trapezoidal rule and the Simpson improvement.

The table above gives the CPU-time needed with the two procedures, and compares the performance using the scalar version TRAPZD and its vector version VTRAPZD.

```
              SUBROUTINE VTRAPZD(FUNC,A,B,S,N)
              PARAMETER (NSCAL=4, NPOINT=2**NSCAL+1)
C WITH N.LE.NSCAL: VECTORS TOO SHORT, SO WORK IN SCALAR MODE
C BUT IN FIRST CALL: ALREADY PREPARE FUNCTION-VALUES FOR ALL
C SCALAR CALLS TO COME
              DIMENSION F(NPOINT)
              EXTERNAL FUNC
              SAVE
              IF (N.LE.NSCAL) THEN
                 IF (N.EQ.1) THEN
                    F(1;NPOINT)=
     =                FUNC(Q8VINTL(A,(B-A)/(NPOINT-1);NPOINT);NPOINT)
                    S=0.5*(B-A)*(F(1)+F(NPOINT))
                    IT=1
                 ELSE
                    IDEL=NPOINT/IT
                    IX=1+IDEL/2
                    SUM=0.
                    DO 11 J=1,IT
                       SUM=SUM+F(IX)
                       IX=IX+IDEL
11                  CONTINUE
                    S=0.5*(S+(B-A)*SUM/IT)
                    IT=2*IT
                 ENDIF
              ELSE
C ADD 2**(N-2) EXTRA POINTS
                 DEL=(B-A)/IT
                 S=0.5*(S+
     +              DEL*Q8SSUM(FUNC(Q8VINTL(A+0.5*DEL,DEL;IT);IT)))
                 IT=2*IT
              ENDIF
              RETURN
              END
```

VECTORIZED Program $\boxed{\text{VTRAPZD}}$

```
              FUNCTION FUNC(X;*)
              DESCRIPTOR X,FUNC
              FUNC=(X**2)*(X**2-2.0)*VSIN(X;FUNC)
              END
```

Vectorized integrand example for VTRAPZD.

V.B. Gaussian Quadrature

In the Gaussian quadratrue rules, one imposes that

$$\int_a^b dx\, W(x)f(x) \approx \sum_{i=1}^N w_i\, f(x_i)$$

Execution time (in sec.) for 14
refinement stages of the trapezoidal rule
for $\int_0^{\pi/2} x^2(x^2-2)\sin(x)\, dx$

With TRAPZD   0.01704   With VTRAPZD   0.0044

```
      SUBROUTINE QTRAP(FUNC,A,B,S)
      PARAMETER (EPS=1.E-6, JMAX=20)
      EXTERNAL FUNC
      OLDS=-1.E30
      DO 11 J=1,JMAX
        CALL TRAPZD(FUNC,A,B,S,J)
        IF (ABS(S-OLDS).LT.EPS*ABS(OLDS)) RETURN
        OLDS=S
11    CONTINUE
      PAUSE 'TOO MANY STEPS.'
      END

      SUBROUTINE QSIMP(FUNC,A,B,S)
      PARAMETER (EPS=1.E-6, JMAX=20)
      EXTERNAL FUNC
      OST=-1.E30
      OS= -1.E30
      DO 11 J=1,JMAX
        CALL TRAPZD(FUNC,A,B,ST,J)
        S=(4.*ST-OST)/3.
        IF (ABS(S-OS).LT.EPS*ABS(OS)) RETURN
        OS=S
        OST=ST
11    CONTINUE
      PAUSE 'TOO MANY STEPS.'
      END
```

Programs $\boxed{\text{QTRAP}}$, $\boxed{\text{QSIMP}}$, from Ref. [4], using TRAPZD.

is exact if $f(x)$ is a polynomial of degree N-1. E.g. if

$$W(x) = \frac{1}{\sqrt{1-x^2}} \quad \text{and } a = -1, \, b = 1,$$

this quadrature formula is called Gauss-Chebychev integration. The most common case is the simple choice $W(x) = 1$ on the interval [-1,1], i.e. the Gauss-Legendre quadrature. The tables of weights and abcissas are found in standard textbooks.

The extension of the rule to an interval [a,b] instead of [-1,1] can be obtained by the change of variables:

$$\xi \in [-1,1] \Leftrightarrow x = \frac{a+b}{2} + \frac{a+b}{2}$$

using:	Execution time (in sec.) for $\int_0^{\pi/2} x^2(x^2-2)\sin(x)\, dx$ QTRAP	QSIMP
With TRAPZD	0.08259	0.00282
With VTRAPZD	0.00258	0.00058

```
          SUBROUTINE QGAUS(FUNC,A,B,SS)
          DIMENSION X(5),W(5)
          DATA X/.1488743389,.4333953941,
     *.6794095682,.8650633666,.9739065285/
          DATA W/.2955242247,.2692667193,
     *.2190863625,.1494513491,.0666713443/
          XM=0.5*(B+A)
          XR=0.5*(B-A)
          SS=0
          DO 11 J=1,5
          DX=XR*X(J)
          SS=SS+W(J)*(FUNC(XM+DX)+FUNC(XM-DX))
11        CONTINUE
          SS=XR*SS
          RETURN
          END
```

Program ⎡QGAUS⎤ from Ref. [4].

The routine QGAUS evaluates the integral $\int_a^b dx\, f(x)$ with a 10-point Gauss-Legendre quadrature according to the rule:

$$\int_a^b dx\, f(x) \approx \sum_{j=1}^{N} \frac{a-b}{2}\ w_j\, f\left(\frac{a+b}{2} + \frac{a-b}{2}\,\xi_j\right)$$

where w_j and ξ_j are the tabulated weights and abcissas on the interval [-1,1].

Although the vectorization is simple, it is normally not worth the effort, because of the limited number of points (if Gauss quadrature is appropriate).

But if the integration is required for a large number of **parameter** values in the function, high performance by vectorization can be obtained **by vectorizing with respect to these parameters.**

Consider e.g. $\int_0^5 dx\, xe^{-cx}$ for 100 values of c, equally distributed between $c_1 = 0.01$ and $c_{100} = 1$ with a vector function VF.

```
      FUNCTION VF(X;*)
      COMMON /PARAM/NC,CC,C(1)
      DESCRIPTOR VF
      ASSIGN VF,.DYN.NC
      VF=X*VEXP(-C(1;NC)*X;NC)
      RETURN
      END
```

Vector function integrand for VQGAUS.

```
      SUBROUTINE VQGAUS(FUNC,A,B,SS)
      DIMENSION X(5),W(5)
      DESCRIPTOR SS
      DATA X/.1488743389,.4333953941,
     *.6794095682,.8650633666,.9739065285/
      DATA W/.2955242247,.2692667193,
     *.2190863625,.1494513491,.0666713443/
      NC=Q8SLEN(SS)
      XM=0.5*(B+A)
      XR=0.5*(B-A)
      SS=0
      DO 11 J=1,5
        DX=XR*X(J)
        SS=SS+W(J)*(FUNC(XM+DX;NC)
     +              +FUNC(XM-DX;NC))
11    CONTINUE
      SS=XR*SS
      RETURN
      END
```

VECTORIZED program ⌈VQGAUS⌉

Execution time (in sec.) with 100
values of the parameter c
for $\int_0^5 dx\ xe^{-cx}$

With QGAUS	0.0210	With VQGAUS	0.0011

With the vector routine VGAUS, the timing table above was obtained as compared with the scalar version QGAUS.

V.C. Chebychev Approximation

Expansion in Chebychev polynomials is the basis of many numerical applications. E.g. the fact that they are orthogonal in the interval [-1,1] over a weight $(1-x^2)^{-1/2}$ makes them very appropriate for differentiation and integration, especially for functions with integrable singularities.

```
        SUBROUTINE CHEBFT(A,B,C,N,FUNC)
        PARAMETER (NMAX=50, PI=3.141592653589793)
        DIMENSION C(N),F(NMAX)
        BMA=0.5*(B-A)
        BPA=0.5*(B+A)
        DO 11 K=1,N
        Y=COS(PI*(K-0.5)/N)
        F(K)=FUNC(Y*BMA+BPA)
11      CONTINUE
        FAC=2./N
        DO 13 J=1,N
        SUM=0.
        DO 12 K=1,N
          SUM=SUM + F(K)*COS((PI*(J-1))*((K-0.5)/N))
12      CONTINUE
        C(J)=FAC*SUM
13      CONTINUE
        RETURN
        END
```

Program ⟨CHEBFT⟩ from Ref. [4].

The Chebychev polynomial $T_n(x)$ of degree n is given by:

$$T_n(x) = \cos(n \arccos x)$$

with recurrence relation:

$$\cos(n\theta) = 2\cos(\theta)\cos([n-1]\theta) - \cos([n-2]\theta)$$

The Chebychev approximation for a function $f(x)$ on the interval $[-1,1]$ assumes that the function can be "accurately" approximated by a polynomial of degree N:

$$f(x) \approx \left[\sum_{k=1}^{N} c_k T_{k-1}(x) - \frac{1}{2} c_1 \right]$$

where the N expansion coefficients are given by:

$$c_j = \frac{2}{N} \sum_{k=1}^{N} f\left[\cos\left(\frac{\pi(k - \frac{1}{2})}{N} \right) \right] \cos\left(\frac{\pi(j - 1)(k - \frac{1}{2})}{N} \right)$$

A change of variables generalizes the integration domain to $[a,b]$:

$$y \equiv \frac{2x - (b+a)}{(b-a)}$$

The routine CHEBFT determines the expansion coefficients for a function on $[a,b]$, and stores them in an array C of dimension N. (In practice $N \approx 50$.) The generalization VCHEBFT of this program for a vector function is also given.

```
         SUBROUTINE VCHEBFT(A,B,C,N,FUNC)
         PARAMETER (PI=3.141592653589793)
         DIMENSION C(N)
         DESCRIPTOR DK,WK,YK,FK
         ASSIGN DK,.DYN.N
         ASSIGN WK,.DYN.N
         ASSIGN YK,.DYN.N
         ASSIGN FK,.DYN.N
         DK=Q8VINTL(0.5*(PI/N),PI/N;N)
         YK=0.5*((B-A)*VCOS(DK;N)+(B+A))
         FK=FUNC(YK;N)
         DO 11 J=1,N
            WK=VCOS((J-1)*DK;N)
            C(J)=Q8SDOT(FK,WK)
11       CONTINUE
         C(1;N)=(2./N)*C(1;N)
         RETURN
         END
```

VECTORIZED program VCHEBFT

Execution time (in sec.) with 40
Chebyshev-expansion coefficients of
$x^2(x^2-2)\sin(x)$ on $[-\pi/2, \pi/2]$

| With CHEBFT | 0.01890 | With VCHEBFT | 0.00177 |

As an example, consider the function $x^2(x^2-2)\sin(x)$ on the interval $[-\pi/2, \pi/2]$. The timing results for calculating the 40 lowest expansion coefficients are given in the table.

The main purpose of the Chebychev approximation of a function is an accurate and fast evaluation in many points. When truncated to $M < N$, the error does not exceed the sum of the neglected c_j's, which allows for an easy estimate of the error.

The evaluation of the function with $M(\leq N)$ expansion coefficients c_j, calculated earlier, is usually realized on the basis of the recurrence relations. But, because of its recursive nature, the procedure is not very appropriate for vectorization.

However, one is usually interested in the approximation for the function in a very large number of points. For this case, a concise implementation of a vector function VCHEBEV OF A VECTOR x is provided, to be compared with its scalar counterpart CHEBEV.

```
              FUNCTION CHEBEV(A,B,C,M,X)
              DIMENSION C(M)
              IF ((X-A)*(X-B).GT.0.) PAUSE 'X not in range.'
              D=0.
              DD=0.
              Y=(2.*X-A-B)/(B-A)
              Y2=2.*Y
              DO 11 J=M,2,-1
                SV=D
                D=Y2*D-DD+C(J)
                DD=SV
11            CONTINUE
              CHEBEV=Y*D-DD+0.5*C(1)
              RETURN
              END
```

Program ⟦CHEBEV⟧ from Ref. [4].

```
              FUNCTION VCHEBEV(A,B,C,M,X;*)
              DIMENSION C(M)
              DESCRIPTOR VCHEBEV,X,TJ,Y
              IF (Q8SCNT((X-A)*(X-B).GT.0).GT.0)PAUSE 'X not in range'
              LENX=Q8SLEN(X)
              ASSIGN TJ,.DYN.LENX
              ASSIGN Y ,.DYN.LENX
              Y=(2*X-(B+A))/(B-A)
              Y=VACOS(Y;LENX)
              VCHEBEV=-0.5*C(1)
              DO 11 J=M,1,-1
                TJ=VCOS((J-1)*Y;LENX)
                VCHEBEV=VCHEBEV+C(J)*TJ
11            CONTINUE
              RETURN
              END
```

VECTOR function ⟦VCHEBEV⟧

The evaluation of the Chebychev approximation for the function $x^2(x^2-2)\sin(x)$ in 50, 500 and 5000 points in the interval $[-\pi/2,\pi/2]$, with 10 out of the 40 expansion coefficients calculated above (to get a relative accuracy of 10^{-6}). The timing table compares the CPU-times.

CONCLUSION

With several general purpose examples, vectorization has been shown to be efficient in less well-known domains, nl. not directly related to matrix manipulations. For most of these examples, the autovectorizers are clearly unable to realize this vectorization, because they do not yet have the required level of artificial intelligence.

In most of the applications, vectorization is first of all a conceptual task, requiring a basic insight in the global structure of a program. But to realize this conceptual vectorization in the actual implementation, it is almost indispensable to have vector

	Execution time (in sec.) for the Chebyshev-approximation with 10 polynomials $x^2(x^2-2)\sin(x)$ on $[-\pi/2,\pi/2]$	
points	With CHEBEV	VCHEBEV
50	0.0015	0.0005
500	0.0151	0.0024
5000	0.1505	0.0218

variables (and some basic routines to handle vectors) available at the FORTRAN level. These tools were at the users disposal in the FORTRAN-200 of CYBER-205 and ETA10, whereas other vendors are reluctant to provide the vector tools in their FORTRAN, because of the transportability problem. But the examples given clearly demonstrate that many programs only are generally believed to be non-vectorizable because of the lack of appropriate vector tools in the FORTRAN. By leaving the vectorization task to the compiler, many vectorization possibilities remain unexploited. We remain convinced that vectors as a variable in the FORTRAN language are absolutely required to realize efficient vectorization, and that the related transportability problems should be overcome by appropriate emulator facilities.

REFERENCES

1. F. Brosens and J.T. Devreese, "Solution of the time-dependent Hartree-Fock equation in the electron gas", in "*Supercomputer Applications*", A.H.L.Emmen, ed., Elsevier, Amsterdam (1985), p. 207-220.
2. F. Brosens and J.T. Devreese, "Frequency-dependent solution of the Hartree-Fock equation in the electron gas", *phys. stat. sol.* (b) 147:173-183 (1988).
3. F. Brosens and J.T. Devreese, "Vectorization techniques and dynamic electron correlations", in "*Scientific Computing on Supercomputers*", J.T. Devreese and P.E. Van Camp, eds., Plenum, New York (1989).
4. W.H. Press, B.P. Flannery, S.A. Teukolsky, and W.T. Vetterling, "*Numerical Recipes: The Art of Scientific Computing*", Cambridge (1986).

ASTRID: A PROGRAMMING ENVIRONMENT FOR SCIENTIFIC APPLICATIONS ON PARALLEL VECTOR COMPUTERS

E. Bonomi, M. Flück, R. Gruber, R. Herbin, S. Merazzi,
T. Richner, V. Schmid and C.T. Tran

GASOV-EPFL, CH - 1015 Lausanne, Switzerland

ABSTRACT

ASTRID is an interdisciplinary project which aims at implementing a set of structured finite element or finite volume programs adapted to the architecture of parallel vector computers with large central memories. All CPU time consuming sections such as mesh generation, matrix construction and solver are highly vectorized and parallelized. Each numerical experiment is an independent program which runs on three virtual machines for preprocessing, computation and postprocessing. One program includes all components required in an advanced numerical experimentation code : the multimachine data base management system MEM-COM to store data and to send it from one machine to the other, man-machine interfaces to set the geometrical and physical input quantities, a structured semi-automatic adaptive mesh generator, the solver and the graphic interpretation system BASPL integrated into MEM-COM. The data management system is very open and can easily be used to connect with any other program which then can take full advantage of the graphics capabilities of BASPL.

1. INTRODUCTION

The purpose of ASTRID is to combine the efforts in the scientific community on the development of software systems for simulation and modeling of different physical phenomena. Each application realized within the ASTRID concept constitutes an independent program using the same system of modules. Conceptually speaking, an ASTRID program runs three virtual machines : the input computer (PC for instance) on which the geometrical and physical data are introduced, the computational unit (such as a supercomputer) on which are executed the modules for the numerical mesh, the construction of the matrix and the right hand side and for the matrix solver and an output machine with sophisticated graphic capabilities. Given a common resolution technique, most parts of various codes will be identical or similar.

The basic idea of ASTRID is that all the computationally intensive parts of simulation programs are adapted to parallel vector computer architectures [1]. Natural parallel processing can be achieved by decomposing the computational domain into subdomains subject to geometrical, physical or computational arguments [1,2]. In the case of a discrete approach such as finite differences, finite elements or finite volumes, a structured mesh [3] is applied to discretize each of the subdomains, leading to structured matrices. In order to avoid matrix structure modifications, boundary conditions are introduced by matrix transformations and by adding dummy variables. The system of linear equations is solved iteratively by a conjugate gradient method with a sub-domain preconditioning or by a multigrid method [4]. Within one subdomain, the construction of the mesh and the matrices as well as the matrix resolution are performed in vector mode. Much effort is done to build large tasks by concentrating most of the computations within each of the subdomains such that it will be possible in the future to run on a parallel machine with distributed memory. One example of an application of this kind is presented in these proceedings [5].

In this paper, we present the programming environment which runs on the input and output computers [6-9]. We describe in detail the interactive man-machine interfaces used to introduce the geometrical (module MiniM) and the physical (module CASE) data and the graphics system BASPL to visualize the results. These modules are activated by means of the ASTRID Command Language. A set of instructions can be put into a script, thus facilitating the work of the user. All these modules and those executed on the supercomputer (MESH and SOLVE) are interconnected through the database management system MEMCOM which guarantees transport of data from one module to the other and from one computer to the other. It can easily be used to connect with any other program which then can take full advantage of the graphics capabilities of BASPL. Examples of programs using the graphics capabilities of the ASTRID system are those described in [10] and [11].

The entire ASTRID system is demonstrated by a practical application [12,13] presented in the last chapter.

2. ORGANIZATION OF ASTRID

2.1. Application modules

The software architecture of ASTRID is illustrated in Figure 1. There are five modules at the application level: **MiniM**odeller (definition of the geometry and mesh generation), **CASE** (definition of the physics of the problem), **MESH** (mesh generation), **SOLVE** (numerical resolution of the problem) and **BASPL** (examination of the results). The modules are independent of each other and the interface between them is done at the level of the database. The design of the interface between modules is thus reduced to the design of data structures [7] which minimize data storage and redundancy while satisfying the data requirements of the application modules. The software system is thus easily extended by introducing applications that read from and write to the database files through the database manager and by adding the necessary data structures to the database definitions.

2.2. Special Characteristics

ASTRID is an integrated toolbox for numerical computation, in particular for the solution of systems of partial differential equations using the finite element method. As such, it provides a user with the tools required to define the problem to be solved, to run the numerical analysis, and to visualize the results of the computation.

There are several important design decisions which distinguish ASTRID from other finite element analysis systems:

Hardware environment

The ASTRID system was created to take full advantage of the power of supercomputers. The computationally heavy part of the processing, i.e., the numerical resolution, is performed on a supercomputer, presently a CRAY-2. The numerical software of ASTRID is vectorized and the model of the geometry chosen (subdomain decomposition and structured grids per subdomain) allows for the use of parallel computers with global memory [1]. A version prepared to run on a distributed memory machine is under construction [14].

The processing which precedes and follows the numerical analysis is performed on a workstation (presently a Silicon Graphics workstation), thus offering the user an interactive interface for the definition of the problem and for the examination of the results.

Subdomain Decomposition

The geometry of the domain on which an analysis is to be run is decomposed into several subdomains [2]. This is done principally for two reasons: first, a division into

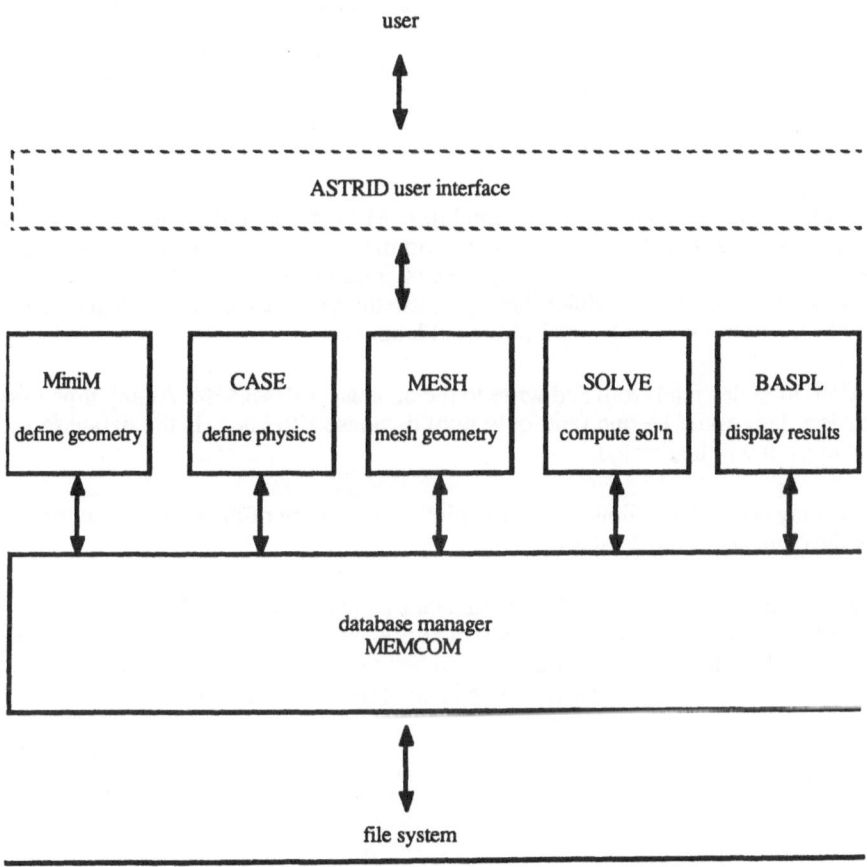

Figure 1. ASTRID software architecture

subdomains assures a better adaptation of the mesh to the geometry of the domain, as each subdomain is meshed separately, second, this decomposition allows for parallelism on the level of subdomains, with calculations on each subdomain being run independently on a separate processor.

The interface between the subdomains is handled using a method described in [4].

Structured Meshing

A structured discretization of a subdomain means that the grid of a subdomain can be mapped onto a cartesian mesh of the unit square or of the unit cube. Such a meshing results in structured matrices, and presents several computational advantages in the optimization procedure for a vectorcomputer. First, it makes for simpler data structures. Second, it provides for easy mesh refinement [3], and lastly, it allows for easier vectorization and parallelization of code [1,4].

Adaptive mesh refinement

The initial mesh of a subdomain is adapted as a function of the solution, thus creating a mesh-solve-mesh cycle which allows us to arrive at a good discretization of the domain [1,3].

3. ASTRID COMMAND LANGUAGE

3.1. User interface

Each of the two pre-processor modules, MiniM and CASE have an **interactive keyword driven interface** which accepts commands specific to the application. Some commands entail subsystems, where a sequence of commands is given that must be terminated by the END keyword. Both modules, however, use the same database and display commands and provide the user with an identical command language.

Both modules read from and write to the database MEMCOM. At any time during the user session there must be one (and only one) database file open. If the database specified does not exist, it will be created.

Display commands allow the user to plot the geometry of the domain and the mesh and to show the contents of the database.

A sequence of MiniM or CASE commands can be written to a file and used as a **script** instead of typing in the commands interactively. A simple **ASTRID command language**, described below, allows the user more flexibility in creating such scripts. A script named file_name is executed by typing @file_name in either MiniM or CASE.

3.2. Command Syntax

Line syntax

Items must be separated by at least one blank. Assignment statements can also use the equal sign, e.g. `DB=FILE.DB` or `DB FILE.DB` are both valid assignment statements. Carriage return or the semicolon (`;`) terminate a command sequence.

Keywords

Keywords are not case sensitive and can be given in upper-case, lower-case or a mixture. They can be abbreviated; the minimum number of characters required depends on the uniqueness of the keyword within the current context and thus differs for each keyword.

Attributes

These consist of integers, integer lists, reals, real lists, strings and text. *Integers* are strings which contain the characters 0-9,+ and - (e.g. 1000, -123,+1). *Reals* are strings which contain the characters 0-9,+,-,.,d/D,e/E (e.g. -1.2345e-7,+.00002,10). **Strings** may contain any combination of characters **except the blank,@,(),[],! and :. Textstrings** may contain any characters and must be delimited by quotes (e.g. COMMX='number of iterations"). **Lists:** must be placed in brackets e.g. [0.0 1.2 3.4 2.4]. Real and integer lists may be generated by [from: to] or using [from:to:step] sequences, e.g. ELEM=[2:5 10 345] assigns the integers 2,3,4,5,10 and 345 to the ELEM key.

Comments

Explanatory text may be placed in scripts on lines beginning with the ! character:

```
!
! select elements
!
SELE NEW BRANCH=1 ALL OFACE=1 END
```

Procedures

These are invoked by giving EXEC=proc_name or by atproc_name. Procedure can have up to 9 parameters and must be referred to within the procedure as $1, $2, $3 etc.

Macro lines

These are text strings that may be invoked by a shortcut command. Up to 9 macro lines may be active. They are defined by the DEFML (define macroline) command:

```
DEFML #i "instructions"
```

where i is in the range of 1 to 9. Macro lines may be invoked by typing #i.

Constants

These are strings that be referenced by their name. They are defined by

```
DEFC varname string
```

Constants are referenced by preceding the constant name with the $ character. They are not typed and their attribute string is simply replaced when the variable name is referenced.

Variables and expressions

Variables and expressions may be assigned, with naming conventions as in FORTRAN. That is, variables or functions whose names start with i,j,k,l,m or n (I,J,K,L,M,N) have integer value ranging from -9999999999 to 9999999999. All other names represent reals. Names can be up to 4 characters long. Note that assignment and evaluation must be places in brackets. Examples:

```
(ic = 5)  (cons = 0.25)  (x=sin(alfa))
```

The command EVAL prints the value of a variable or an expression. Arguments to functions may not have the same name when they are defined and when they are referenced. The following sequence outputs the number 5:

```
(f(x,y)=x*x+y*y)  (a=1)  (b=2)  (EVAL f(a,b))
```

Control statements

A limited number of control statements are available, such as LOOP, BREAK and IF. The following example creates a series of 10 points using the functions x(t),y(t) and z(t). For each iteration of the loop the command CP is executed with a new value for the point.

```
(x(t)=t)  (y(t)=t/2)  (z(t)=t*2)  (i=1)
LOOP (BREAK (i>10) CP (i) (x(i)) (y(i)) (z(i)) (i=i+1))
```

LOOP executes the statements inside the brackets until a BREAK is encountered or until the maximum loop count is reached (see setting interpreter parameters).

The syntax of the IF statement is IF (condition) (expression). Example:

```
IF (a>9)  (EVAL f(a,b))
```

If the condition is satisfied then all the commands in the brackets are executed.

Scripts

A sequence of MiniM or CASE commands can be written to a file and used as a script instead of typing in the commands interactively. The simple command language described above allows the user more flexibility in creating such scripts. A script named file_name is executed by typing **@file_name** or EXEC=file_name.

3.3. Database commands

Both modules, MiniM and CASE, read from and write to a database. At any time during the user session there must be one (and only one) database file open. If the database specified does not exist, it will be created.

Open the database	`DB=file_name`
Close the database	`CLOSE`
Save all changes to the database and continue	`SAVE`
Quit the module and save changes	`QUIT`

4. MINIM: MINI-MODELLER TO DEFINE THE GEOMETRY

In the **MiniM**odeller module, data must be entered which allows for a modelling of the domain which is sufficient for obtaining a good discretization. It is at this level that the user specifies the decomposition into subdomains. Each subdomain is then defined as being composed of 6 faces (3D) or 1 face (2D). Each face can be defined either as a regular grid of points (bilinear patch) or, in the case of a plane surface, as delimited by 4 sides. Each line is in turn defined as a sequence of points, and each point is defined by a set of coordinates (x,y,z). Each 3D subdomain is either defined by a set of coordinates or by a sequence of surfaces.

4.1. Create database objects

..Create point   `CP ip x y z ia`

The integer ip is the index of the created point and must be unique, x, y and z are the coordinates in the three directions and ia is an attribute which can be given to the point.

..Create face

```
CF
        FACE=if1                    face index
        TYPE=BLP                    Bilinear patch face
        LS1=ix                      number of CAD points in x direction
        LS2=iy                      number of CAD points in y direction
        CP=[ip1 ip2...ipN]          gives the points defining the bilinear patch.
                                    N=ix*iy
STORE
END
```

Note: several faces can be defined using the same CF command, e.g.:

```
CF
        FACE=if1   ...face parameters...
STORE
        FACE=if2   ...face parameters...
STORE
        etc.
END
```

The same holds true for the CSD command. Note that the indentation given above is only for purposed of readability, all the keywords of the command can in fact appear on the same line.

..Create subdomain

An object is subdivided in subdomains. This subdivision is done following geometrical, physical and computer architectural arguments. The geometrical decomposition has to be done in such a way that each subdomain can be considered to be a deformed cube which can be meshed in a structured and numerically acceptable way.

```
CSD
        SUBD=isd             index of subdomain
        ISX=iside_x          side index of side x, ISX=1 for 2d, 1...6 for 3d.
STORE
END
```

..Create structured subdomains

```
CSSD
      NSD1=nsd1                        number of subdomains in direction x
      NSD2=nsd2                        number of subdomains in direction y
      NSD3=nsd3                        number of subdomains in direction z for 3D
      NCP1=[in1 in2 in3 ... insd1]     number of CAD points per SD in x
      NCP2=[jn1 jn2 jn3 ... jnsd2]     number of CAD points per SD in y
      NCP3=[kn1 kn2 kn3 ... knsd3]     number of CAD points per SD in z END
```

4.2. Modify data base objects

..Modify point   `MP ip x y z 0`

..Modify face

```
MF
      FACE=if1                         face index
      LS1=ix                           number of CAD points in x direction
      LS2=iy                           number of CAD points in y direction
      CP=[ip1 ip2...ipN]               gives the points defining the bilinear patch.
                                       N=ix*iy
STORE
END
```

..Modify subdomain

```
MSD
      SUBD=isd                         index of subdomain
      ISX=iside_x                      side index of side x, ISX=1 for 2d, 1...6 for 3d.
STORE
END
```

(4.3) Remove database objects

..Remove point   `RP ip`
                            `RP [ip:jp]`   from point ip to point jp
                            `RP [ip:jp:kp]`   as above but in steps of kp

..Remove face   `RF iface`

..Remove subdomain   `RSD isub`

5. CASE: INTERFACE TO DEFINE PHYSICAL QUANTITIES

Once the geometry has been defined, it serves as a base for the introduction of information which describes the problem to be solved. In module **CASE**, boundary and initial conditions are specified, material properties are defined, as well as degrees of freedom, fixed or variable right hand sides. It is here that the user specifies which solver is to be used.

5.1. Analysis directives

This section of definition of the physical quantities starts with the `ADIR` command and, as all partial scripts, ends with `END`.

```
ADIR
        NDIM=n                  Dimension of the problem (n=1,2 or 3)
        ANALYSIS=s              Type of operator:     s=LAPLACE
                                                      s=SOLIDIF
                                                      s=STOKES
                                                      s=NAVIER
        DISCRET=s               Discretization:       s=Q1   Q1 finite elements
                                                      s=H    hybrid elements
        NDF=n                   Number of variables per node
        HEAD='...'              Problem title
END
```

5.2. Boundary conditions

This section starts with `SUBD` and ends with `END`.

```
SUBD
        ISD=isd                 Subdomain number
        STORE                   Makes changes permanent
        TYPE=s                  s=A             subdomain exists
                                s=D             subdomain is dummy
        EPROP=n                 Material property index for subdomain isd
        BCFn i j                Defines boundary condition   n: surface label (from 1 to 6)
                                                             i: variable to be constrained
                                                             j: boundary condition code
END
```

5.3. Material constants

The material properties are set between `EPROP` and `END`.

```
EPROP
        MAT=id                  Identifies material
        NEW                     Initialize new material
        STORE                   Makes definitions or modifications permanent
        TYPE=s                  Type: s=CONST    homogenuous in space and time
                                      s=UDEF     defined by user
        IDENT=n                 If UDEF, calls subroutine EPRPn
        VAL=r                   If CONST, sets value r
END
```

6. MESH : NUMERICAL MESH

6.1. Autoadaptive mesh

As mentioned before, the technique applied in ASTRID to get parallelism in a natural way is to decompose a domain into subdomains [1,2]. Each subdomain is handled by a separate virtual processor. Interprocessor communication is minimized in order to be efficient when the target machine is a parallel architecture with distributed memory.

Each subdomain is considered to be homeomorphic to a cube. This makes it possible to create a structured mesh. A mesh density function which reflects the numerical error of the physical solution is used for grid adaptation [3]. This can be done in an iterative form: generate the mesh, compute the physical solution, choose a new mesh density function after analysis of the numerical error of the physical solution, generate the new mesh by redistributing the grid points in accordance with the mesh density function, recompute the physical solution, and so on. As the number of mesh points and the mesh density function are the same on the interfaces between the subdomains, the resulting grids are continuous.

An example of an adaptive mesh published in [3] is shown in Figure 2.

6.2. Mesh one subdomain

Each subdomain is given by a structured geometrical mesh in the three-dimensional space consisting of a certain number of surfaces.

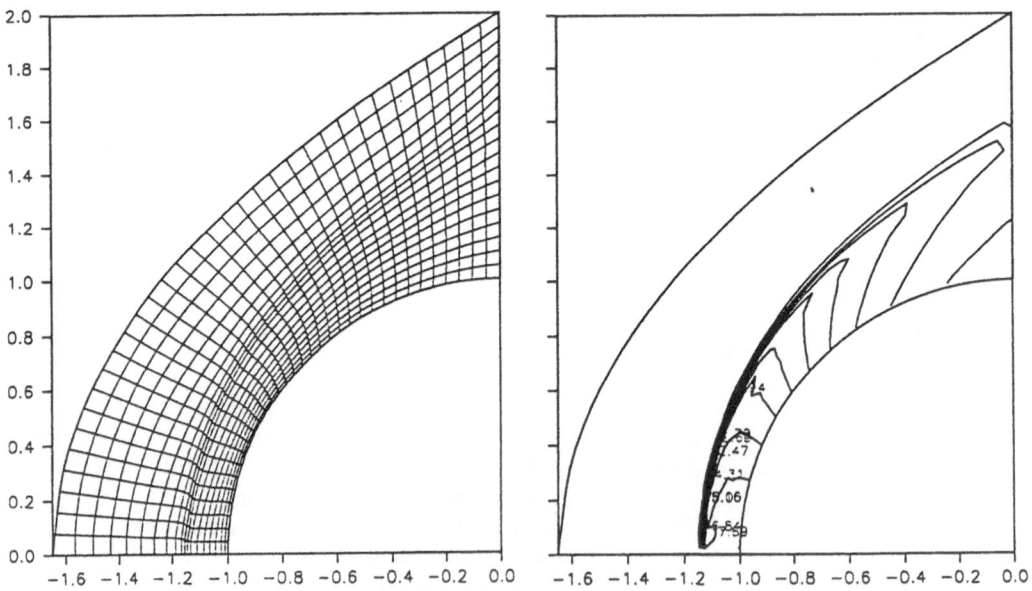

Figure 2. Shock capturing of an Euler flow over a sphere. The adapted grid and the resulting pressure contours.

```
CM
      ISD=isd                      the subdomain index
      MESH=Q                       type of mesh: quadrilateral
            H                      type of mesh: hexagonal
      TYPE=PLN                     type of surface: planar
            CRV                    type of surface: curved
      N1=n1                        number of mesh cells in x direction
      N2=n2                        number of mesh cells in y direction
      N3=n3                        number of mesh cells in z direction
END
```

Note: The TYPE attribute selects the meshing algorithm to be used. PLN selects a meshing algorithm based on molecular dynamic which can be used only for planar surfaces. This type demands that the subdomain be not too concave. CRV selects a meshing algorithm that gives a Coons representation of the surface and can be used for either planar or curved surfaces.

6.3. Mesh all subdomains

It is possible to mesh all subdomains at once:

```
MSSD
      MESH=Q                       type of mesh: quadrilateral
            H                      type of mesh: hexagonal
      TYPE=PLN                     type of surface: planar
            CRV                    type of surface: curved
      NSD1=n1                      number of subdomains in x direction
      NSD2=n2                      number of subdomains in y direction
      NSD3=n3                      number of subdomains in z direction
      N1=[in1 in2 in3 ... insd1]     number of CAD points per SD in x
      N2=[jn1 jn2 jn3 ... jnsd2]     number of CAD points per SD in y
      N3=[kn1 kn2 kn3 ... knsd3]     number of CAD points per SD in z END
```

Note: This instruction can only be used to mesh objects that have been created using the CSSD command.

7. SOLVE: SOLVES THE PROBLEM

The major steps in the solver module [4] are the construction of the matrix and the right hand side vector as well as the matrix solver which can be based on a direct method (Gauss elimination technique) or an iterative method (Conjugate gradients with preconditioning or Multigrid). We very briefly describe these modules.

7.1. Construction of the matrix and right hand side

The contributions to the matrices and to the right hand sides of each subdomain are all constructed in parallel. This is possible since mesh points belonging to the interfaces between subdomains are duplicated [4].

The boundary conditions are imposed in such a way that the structured matrix form is untouched. The variables to be eliminated are replaced by dummy ones. The real matrix problem then consists of the ensemble of all partial problems for each subdomain plus all the connectivity conditions to be applied on the variables of the interfaces.

We have to mention here that the construction of the submatrices and the subvectors are done in such a way that full vectorization is achieved, the innermost DO-loop going over all mesh cells of the subdomain.

The command to construct the matrix and the right hand side is MVG. The boundary conditions are imposed by IMBC.

7.2. Direct matrix solver

In the case of a direct solver, it is necessary to introduce the connectivity conditions into the final matrix problem

$$Ax = b \tag{1}$$

In the case with 3 subdomains, this problem (1) writes:

$$
\begin{bmatrix}
\tilde{A}_1 & & & B_1 \\
& \tilde{A}_2 & & B_2 \\
& & \tilde{A}_3 & B_3 \\
C_1 & C_2 & C_3 & A_0
\end{bmatrix}
\begin{bmatrix}
\tilde{x}_1 \\
\tilde{x}_2 \\
\tilde{x}_3 \\
x_0
\end{bmatrix}
=
\begin{bmatrix}
\tilde{b}_1 \\
\tilde{b}_2 \\
\tilde{b}_3 \\
b_0
\end{bmatrix}
\tag{2}
$$

The vector $x = (\tilde{x}_1, \tilde{x}_2, \tilde{x}_3, x_0)$ is composed of the vectors $\tilde{x}_1$, $\tilde{x}_2$ and $\tilde{x}_3$ including the variables which do not lie on common interfaces and of the connectivity vector x_0.

Matrix problem (2) is solved by a block Gauss elimination procedure which writes (i=1,2,3):

$$A_0 = A_0 - \sum C_i \tilde{A}_i^{-1} B_i$$

$$b_0 = b_0 - \sum C_i \tilde{A}_i^{-1} \tilde{b}_i \tag{3}$$

The operations in the sum can be performed in parallel. Inverting $\tilde{A}_i$ means decomposing it in two (lower and upper) triangular matrices $\tilde{A}_i = L_i R_i$ and solving $L_i v_i = c_i$ and $R_i u_i = v_i$

where c_i are colomn vectors of the matrix B_i or the $c_i = \tilde{b}_i$. The vectors u_i are the colomn vectors of the resulting $\tilde{A}_i^{-1} B_i$ matrix or $u_i = \tilde{A}_i^{-1} \tilde{b}_i$.

The remaining matrix problem is $A_0 x_0 = b_0$, where A_0 is full. After having solved for x_0, the vectors $\tilde{x}_i$ are obtained by the back substitution $\tilde{x}_i = \tilde{A}_i^{-1} (\tilde{b}_i - B_i x_0)$.

Let us note here that the block Gauss elimination procedure (3) is performed on sparse matrices $\tilde{A}_i$, B_i and C_i. If the number of unknowns in $\tilde{x}_1$, $\tilde{x}_2$ and $\tilde{x}_3$ is related to an n-dimensional problem, the connectivity vector x_0 has a number of unknowns related to a (n - 1) dimensional problem. If n = 2 this procedure can be very efficient. If n = 3, the number of variables in the two-dimensional connectivity problem becomes very large in most of the practical cases. Iterative matrix solvers such as multigrid or conjugent gradient methods should rather be used.

The instruction to resolve the matrix problem by the direct solver is LU.

7.3. Iterative matrix solvers

To solve matrix problems coming from a finite element, finite volume or finite difference approach of a realistic three-dimensional numerical experiment, iterative methods have to be chosen [15]. This is the only way to preserve the sparsity pattern of matrix A in equation (2). These iterative solvers can be of type Jacobi, Gauss-Seidel with over- or under-relaxation, Conjugate Gradient method with or without preconditionning, multigrid or a pseudo-time evolution. The most time-consuming steps in these methods are generally the matrix-vector multiply procedure. It is exactly this operation which in our approach has been optimized to reach a high level of efficiency in the use of modern parallel vector computers such as CRAY 2. The matrix A of the matrix-vector multiply $u = Ay$ is given by the submatrices A_i and by the connectivity conditions.The vector y reflects an estimate or a correction of x. In contrast to the direct solvers, the connectivity conditions are not applied to the submatrices A_i directly, but on the resulting vector u.by:

. Multiply all unassembled submatrices A_i with the subvectors y_i which include the boundary conditions but not the connectivity conditions. The results are the unassembled vectors u_i.

. For all connection points, different values of u are obtained in different subdomains. The actual values at these points are the sum of the different components of u corresponding to the same mesh point.

. Each of the summed up values has to be redistributed to all the subdomains to which the connection points belong.

The instructions to give for iterative solvers are CG and MG (see[4]).

8. BASPL: GRAPHICS SYSTEM

8.1. Fundamental remarks

BASPL [8] is a post-processing graphic system for finite element, finite volume and finite difference applications. Its data structure is based on the one defined by MEMCOM [6]. This implies that BASPL can be used for graphical representation of any data conforming

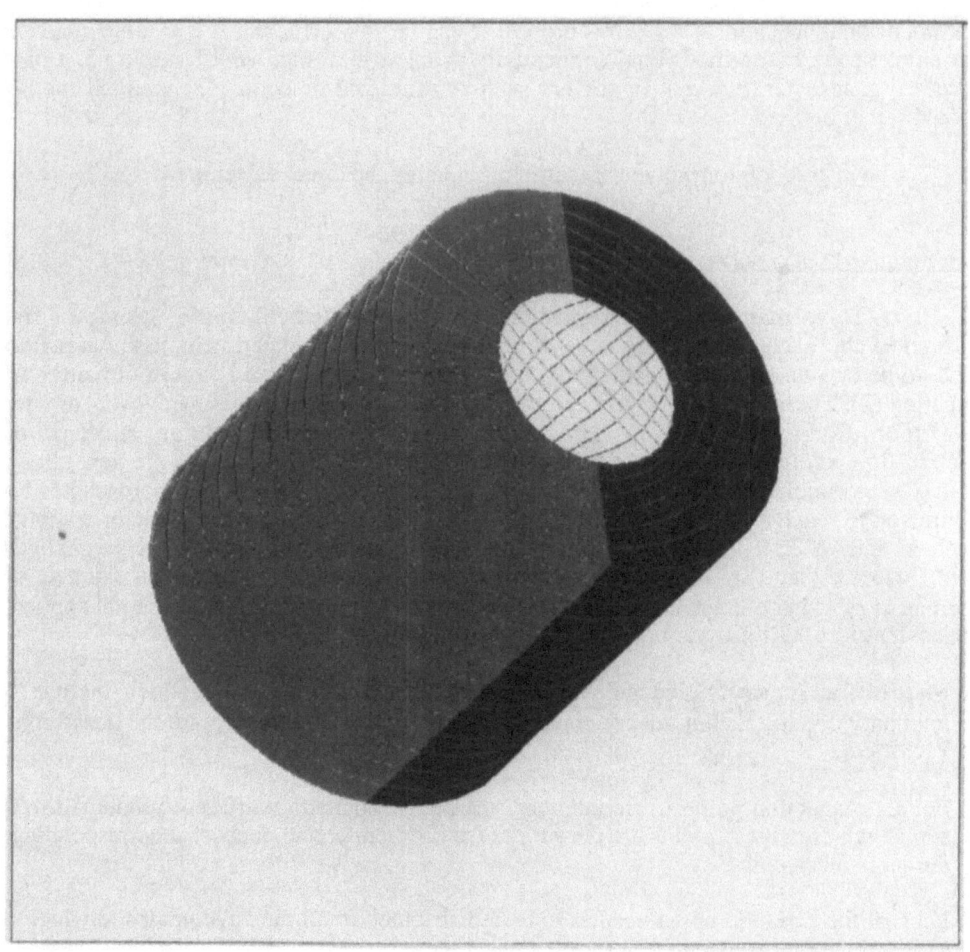

Figure 3a

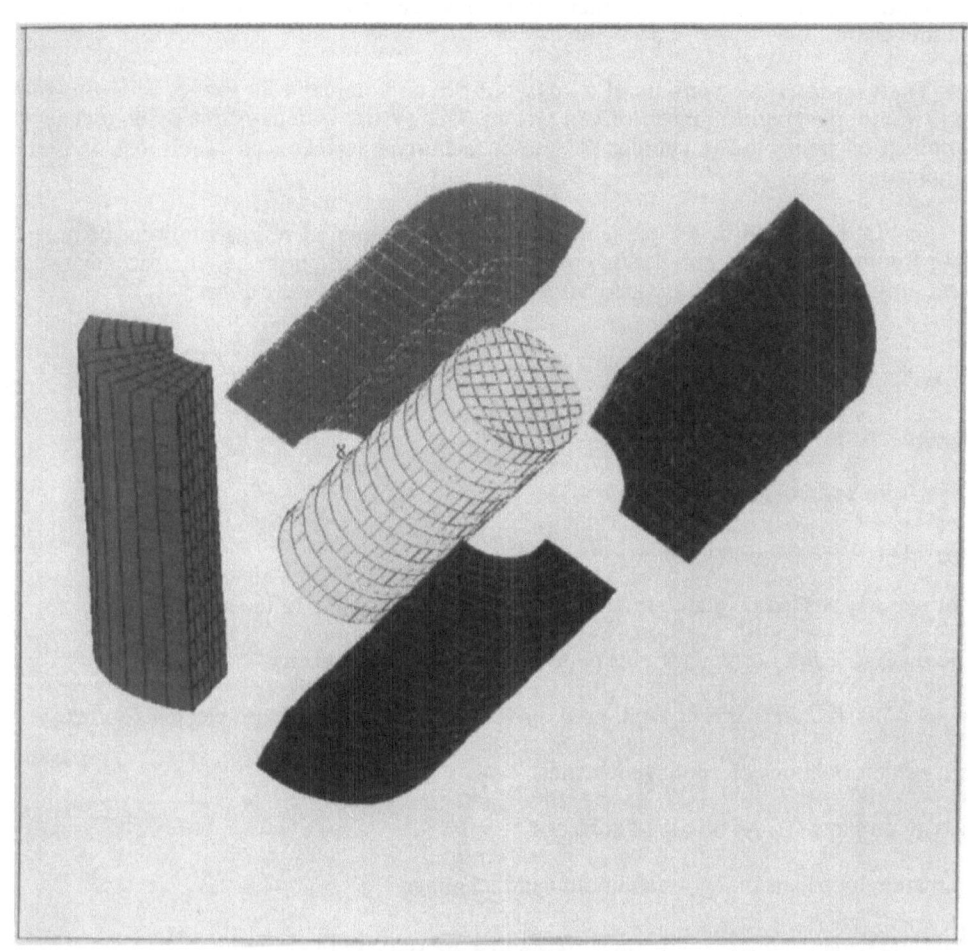

Figure 3b

MEMCOM. If a user would like to take advantage of this software, he has to bring his data in MEMCOM format which for finite element or finite volume problems implies to write a simple interface. This step has already been made for several programs coming from fluid mechanics, material science, physics and mathematics. Examples are those described in [5,10,11].

These programs can be based on a structured or unstructured grid, both approaches can be handled by BASPL. Also, a domain can be cut into subdomains as shown in Figure 3.

The BASPL program has been optimized to run on very fast high-end graphic workstations. At the moment, implementations has been realized on Silicon Graphics and Hewlett Packard.

The user interface consists of a keyword-driven command language. The user may swap to the simple graphical menu-driven system. This system is demonstrated by means of the application presented in chapter 9. A user-friendly interface à la MacIntosh is under development.

In Figures 4 and 5, we present, as examples, graphical representations of results coming from two different simulation codes [10,11]. For these purposes, interfaces between the user programs and the well-defined MEMCOM data structure were done.

8.2. Functionalities of BASPL

.. Diagrams f(x)

.. 3D surfaces and volumes by elements

.. Plane intersections through 2D and 3D objects

.. Continuously shaded colour contour plots of scalar fields on top of the selected geometry

.. Colour contour lines of scalar fields on top of the selected geometry

.. Vector plots, monochrome and coloured, of vector fields on top of the selected geometry

.. Plots of deformations of initial geometries

.. "Mountain plots" f(x,y) on top of surfaces

.. Animation by means of a powerful command language

.. Static and dynamic display of particle tracks in a 3D velocity field

.. Different diagnostics items

9. APPLICATION: DISTRIBUTION OF ELECTRICAL CONTACTS

9.1. The physical problem

The framework of this work is the mathematical modellisation of Solid Oxide Fuel Cells (SOFC) [12]. The problem which we are interested in is the minimisation of ohmic losses in the electrodes and electrolyte. Indeed, an important problem in the conception of a SOFC is to get the maximum use of the electric current which is generated by the oxygen ions diffusion. We describe here a simplified model of this problem:

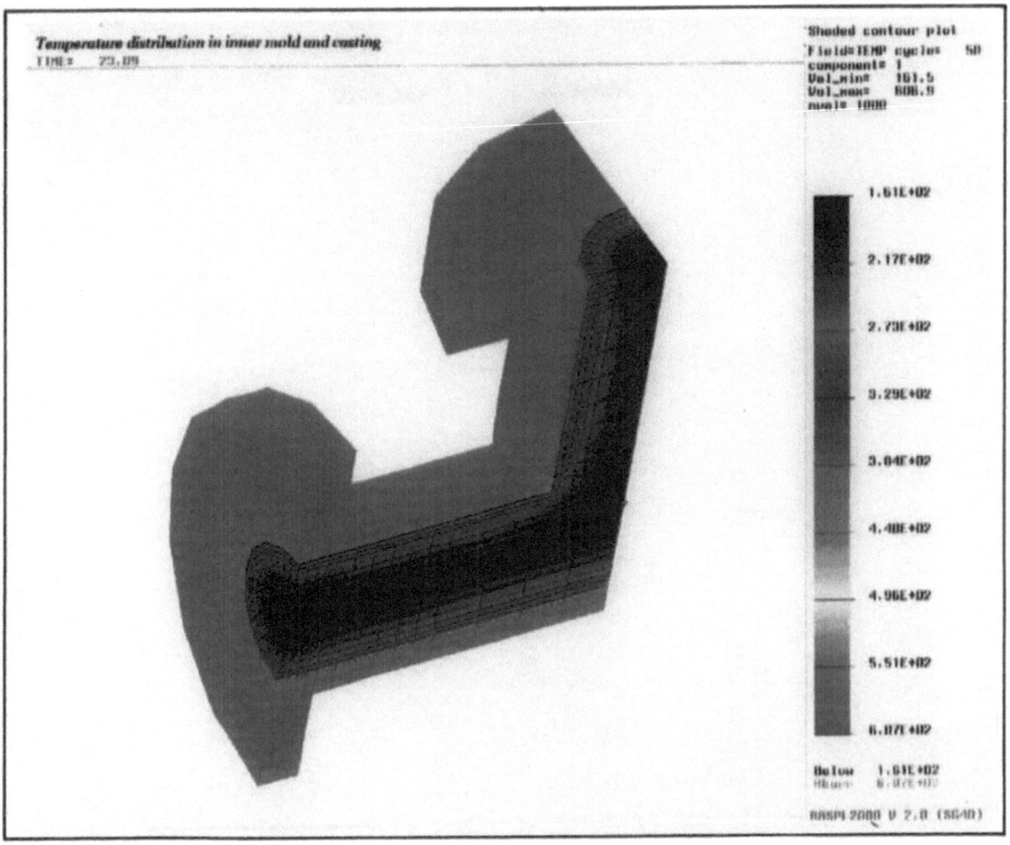

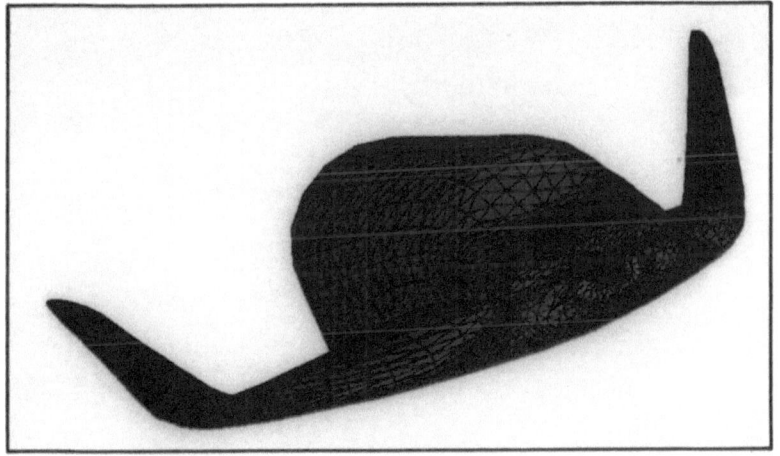

Temperature distribution in inner mold and casting
TIME = 73.89

Shaded contour plot
Field=TEMP cycles 50
components 1
Val_min= 161.5
Val_max= 606.9
axis 1000

1.61E+02

2.17E+02

2.73E+02

3.29E+02

3.84E+02

4.40E+02

4.96E+02

5.51E+02

6.07E+02

Below 1.61E+02
Above 6.07E+02

RASP 2000 V 2.0 (SG4D)

Figure 4

Figure 5

Consider a rectangular thin plate made of a conductive material. On one of its faces, the electric potential is set to zero; on the other face, it is set to one on the contact area. The other sides of the plate are electrically insulated (see Figure 6). Our aim is to compute the electric current which flows through the plate, due to the forced potential. In the real case of an electrolyte-electrodes plate of a SOFC, the electric potential is generated by the oxygen ions flux, and not externally imposed; moreover, the three constituents (i.e. anode, electrolyte and cathode) do not have the same electric conductivities. However, our first simple model will help understanding how the contacts to the electrodes should be set so that the cell generates the maximum current. Let us now turn to the mathematical model.

Let $\Omega \subset \mathbb{R}^3$ be the parallelepipedon defined by $\Omega = \{ (x,y,z) \mid 0 < x < L, 0 < y < L, 0 < z < d \}$, where $L = 0.05$ m and $d = 0.0025$ m. Let Γ_0 be the face of Ω defined by $z = 0$, Γ_1 the part of the face defined by $z = d$ where the potential is set to 1, and $\Gamma = \partial \Omega \setminus (\Gamma_0, \Gamma_1)$ the remainder of the boundary of Ω, which is insulated as shown on Figure 6.

Let us define the following problem:

Find an electric potential $\phi : \Omega \rightarrow \mathbb{R}$ such that

$$-\Delta \phi = 0, \quad \text{in } \Omega, \tag{4}$$

$$\phi = 0, \quad \text{on } \Gamma_0, \tag{5}$$

$$\phi = 1, \quad \text{on } \Gamma_1, \tag{6}$$

$$\frac{\partial \phi}{\partial n} = 0, \quad \text{on } \Gamma. \tag{7}$$

Equation (4) is the Laplace equation satisfied by the potential in the plate. Equations (5) and (6) give the potential which is imposed on the bottom and top faces of the plate. Equation (7) states the electrical insulation of the part Γ of the boundary.

It can be easily shown that this problem has a unique solution ϕ, in a mathematical sense that we shall omit here. The total current flowing through the parallelepipedon is then given by the following formula:

$$I_{\Gamma_1} = \int_{\Gamma_1} \nabla \phi \cdot \mathbf{n} \, ds, \tag{8}$$

where $\mathbf{n}$ is the unit outward normal vector to Ω and $\nabla \phi$ is the gradient of ϕ,

i.e. $\nabla \phi = (\frac{\partial \phi}{\partial x}, \frac{\partial \phi}{\partial y}, \frac{\partial \phi}{\partial z})^T$.

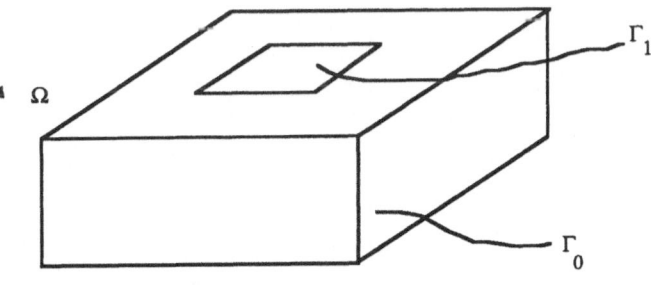

Figure 6. Geometry of the problem

The geometrical dimensions of Ω have been derived from the study of a real cell. The conductivity and the imposed potential difference have been chosen equal to 1, since the problem is linear.

We solve the problem by first describing the geometry with MiniM and the physical quantities with CASE, then we mesh with MESH and solve the problem with SOLVE. Finally, the postprocessing is done using BASPL.

9.2. MiniM

```
! problem of electric plate: Minim script
!
db=elect.db
!
! define parameters
!
(a=0.05)
(d=a*a)
(delt=0.1*a)
(cent=0.5*a)
(alow=cent-delt)
(high=cent+delt)
!
! define the points
!
CP 1 0 0 0 0
CP 2 (alow) 0 0 0
CP 3 (high) 0 0 0
CP 4 (a) 0 0 0
          .
CP 29 0 (a) (d) 0
CP 30 (alow) (a) (d) 0
CP 31 (high) (a) (d) 0
CP 32 (a) (a) (d) 0
!
! define the subdomains
!
CSSD
  nsd1=3
  nsd2=3
  nsd3=1
  ncp1= [2 2 2]
  ncp2= [2 2 2]
  ncp3= 1
  STORE
END
!
! mesh all the subdomains
!
MSSD
  nsd1=3
  nsd2=3
  nsd3=1
```

```
     n1= [4 2 4]
     n2= [4 2 4]
     n3= 10
     STORE
END
!
! plot the mesh and points
!
set nat n on
plot mp
```

<u>9.3. CASE</u>

```
! problem of electric plate: Case script
!
db= elect.db
!
! define type of problem (analysis directives)
!
ADIR
  ndim=3
  analysis='LAPLACE'
  discret=Q1
  prec=E
  ndf=1
  head='Electric Plate'
END
!
!  subdomain boundary conditions
!
SUBDM
   isd=5
   eprop=1
! b.c. subdomain=5 face=2 dof=1 constant=1
   bcf2 1 1 1.0
   store
   (i=1)
! b.c. face=1 dof=1 constant=0
! for all subdomains
   loop (break (i>9) isd=(i) eprop=1 bcf1 1 1 0. store (i=i+1))
  END
!
! define conductivity constant
!
EPROP
  mat=1
  type=const
  val=1.0
END
```

<u>9.4. SOLVE</u>

The solution ϕ to the above problem cannot be obtained explicitly for any Γ_1. Our aim here is to compute an approximation to ϕ by the finite element method. First we define a mesh

Figure 7

of the domain Ω, i.e. a partition into a finite number of disjoint parts (elements) which we will choose to be hexaedra, so as to be compatible with the ASTRID data structure [7]. We then seek an approximate potential with a known dependency on x, y and z on each element (in our case, linear in x, y and z) and which is continuous on Ω. Figure 7 gives a domain decomposition of Ω and its meshing in hexaedra. Equations (4) to (7) are then transformed into their so-called weak form. The approximation of the solution and of the test functions leads to a regular system of linear equations which is solved by an adequate numerical method, programmed on a computer. The finer the mesh is, the more accurate will the computed solution be. Unfortunately, the finer the mesh is, the larger the linear system is (and therefore the computer cost).

The instruction set is:

```
! problem of electric plate:SOLVE script
!
db=elect.db
SOLVE
! create matrix and right hand side
        MVG
! impose boundary conditions
        IMBC
! solve by conjugate gradients
        CG 100 1.e-5
END
```

9.5. Numerical results

We present here results which were obtained using a mesh of Ω consisting of N*N*N rectangular parallelepipedons, with N = 10, 20, 30 and 40 (see Figure 8).

Two different shapes of Γ_1 (Γ_1^a and Γ_1^b) were studied:

- Γ_1^a is the square defined by: 0.4 L<x<0.6 L , 0.4 L<y<0.6 L , z=d.

- Γ_1^b consists in the four squares defined by:

    0.2 L<x<0.3 L , 0.2 L<y<0.3 L , z=d,
    0.2 L<x<0.3 L , 0.7 L<y<0.8 L , z=d,
    0.7 L<x<0.8 L , 0.2 L<y<0.3 L , z=d,
    0.7 L<x<0.8 L , 0.7 L<y<0.8 L , z=d.

The surfaces Γ_1^a and Γ_1^b have the same area, but with a different setting on the face z=d. In both cases, the potential ϕ was computed by the finite element method, and the current deducted by formula (8). The linear system is solved by the conjugate gradient method with diagonal preconditioning. Figure 8 shows the value of the computed current I_{Γ_1} according to the size h=L/N of the mesh.

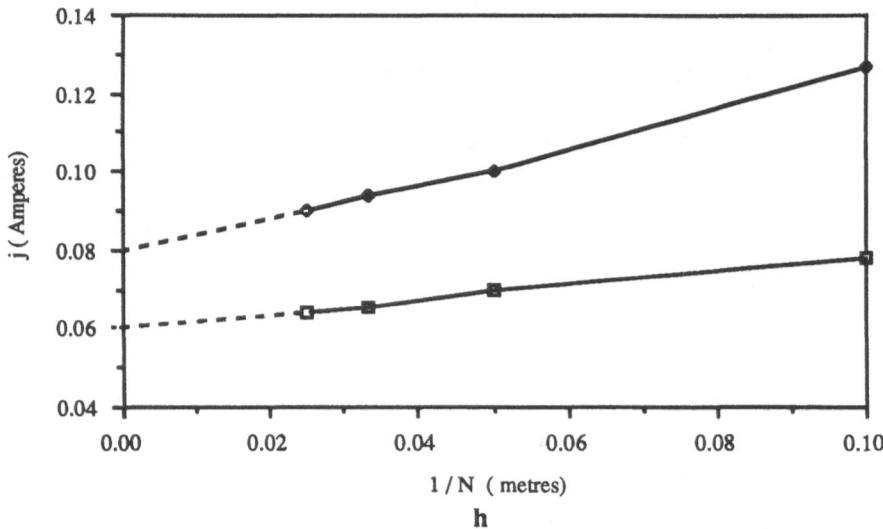

Figure 8. Convergence diagram of the current $j=I_{\Gamma_1}$ according to h=L/N. Empty squares : case of Γ_1^a. Full squares : case of Γ_1^b .

Note that the current converges linearly to the value 0.06 Amperes in the case where $\Gamma_1=\Gamma_1^a$, and to 0.08 Amperes in the case where $\Gamma_1=\Gamma_1^b$. Hence, for the same contact area, the current which flows through the plate is larger when the contact surface is made of four separate parts.

Figure 9 shows the results obtained for Γ_1^a. The equipotentials are shown on the slice x = 0.5 L of the domain (shade representation on Figure 9a and line representation on Figure 9b). Figure 9c represents the equipotentials on the slice z=0.5 d of the domain. The computed total current through the plate is I=0.062 A .

Figure 10 shows the results obtained for Γ_1^b. The equipotentials are shown on the slice x=0.25 L (shade representation on Figure 10a and line representation on Figure 10b). Figure 10c represents the equipotentials on the slice z=0.5 d of the domain. The computed total current through the plate is I=0.090 A , and is thus larger than in the above case, for the same contact area.

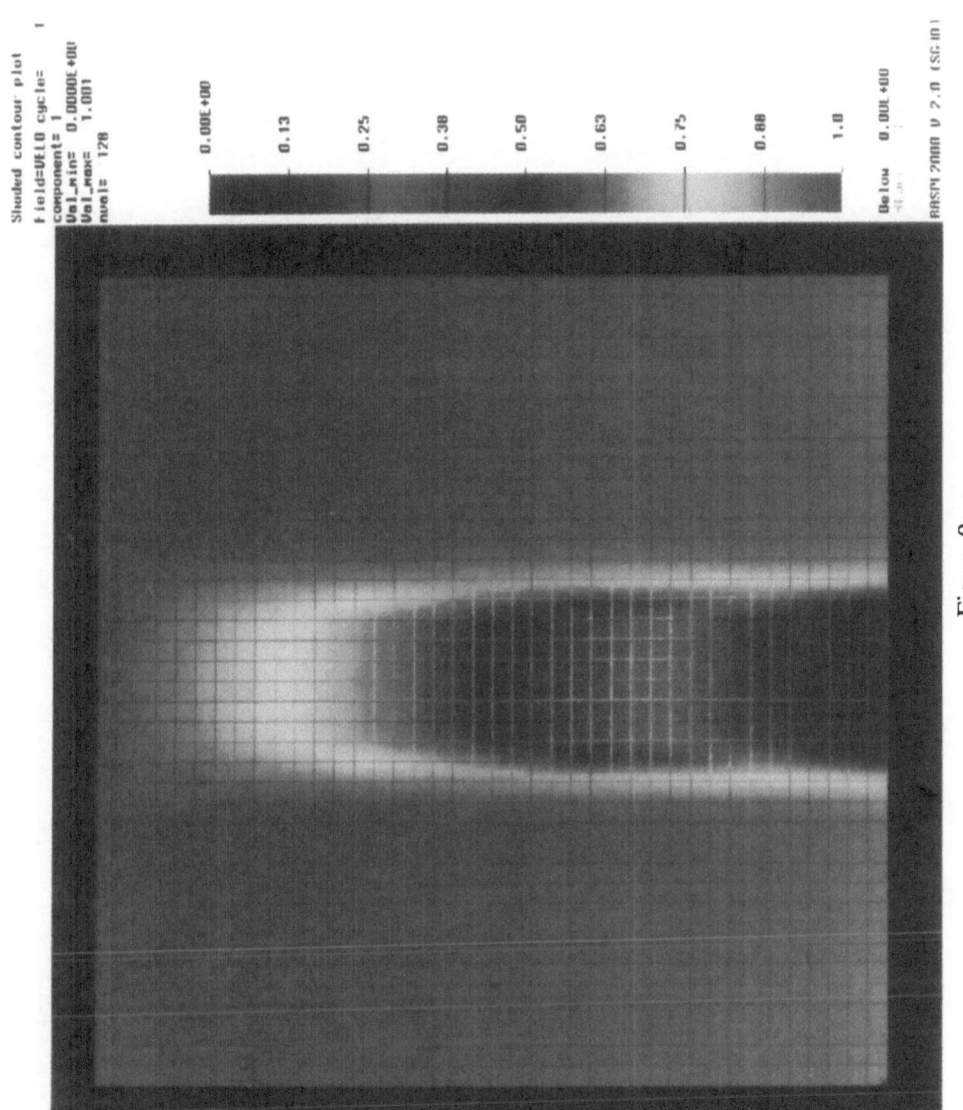

Figure 9a

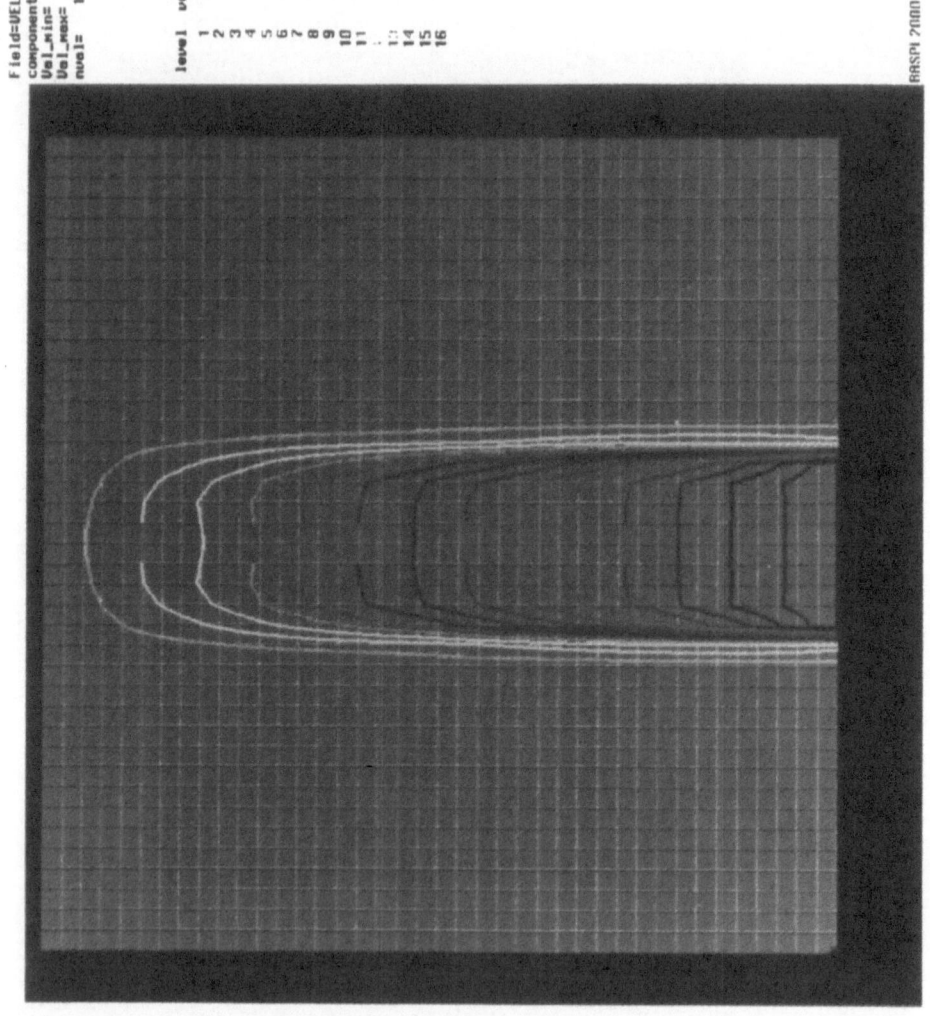

Figure 9b

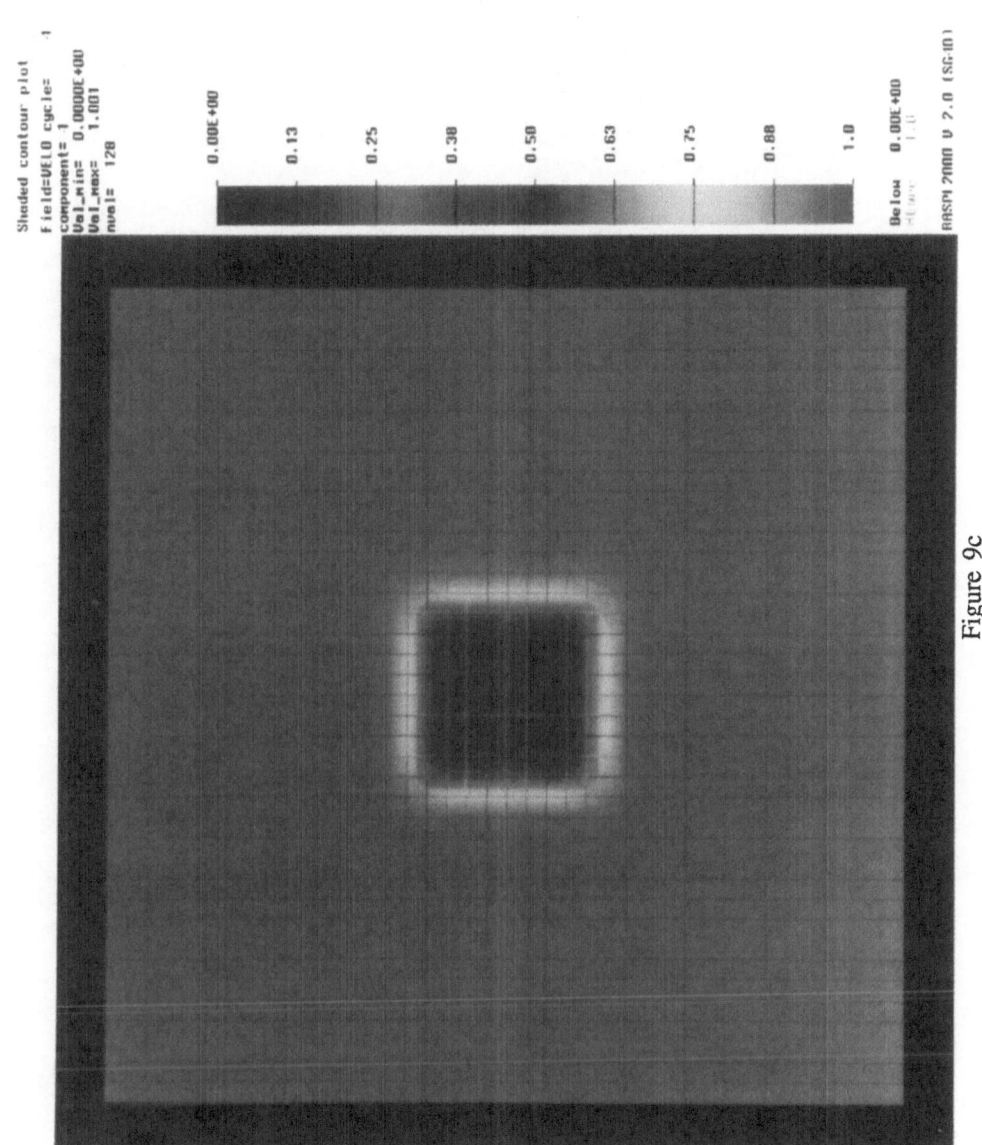

Figure 9c

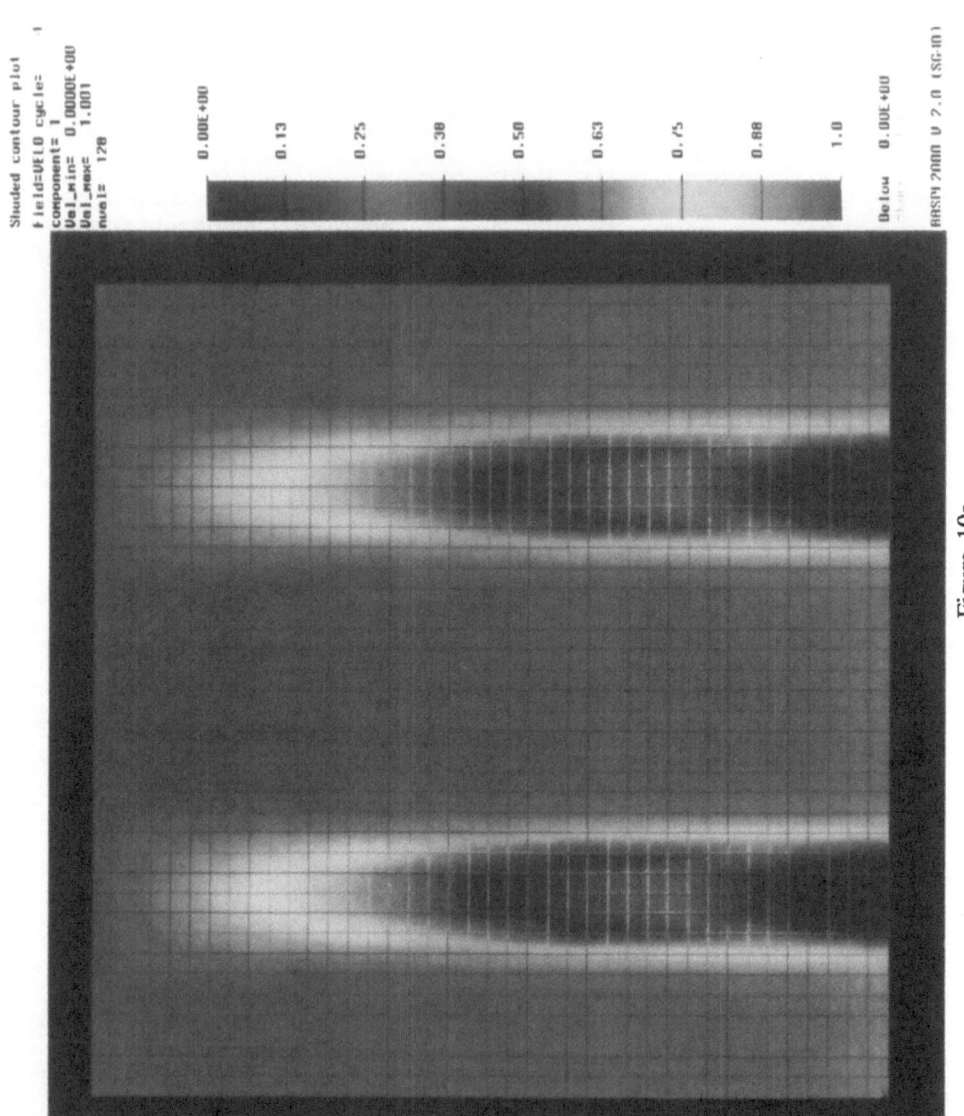

Figure 10a

78

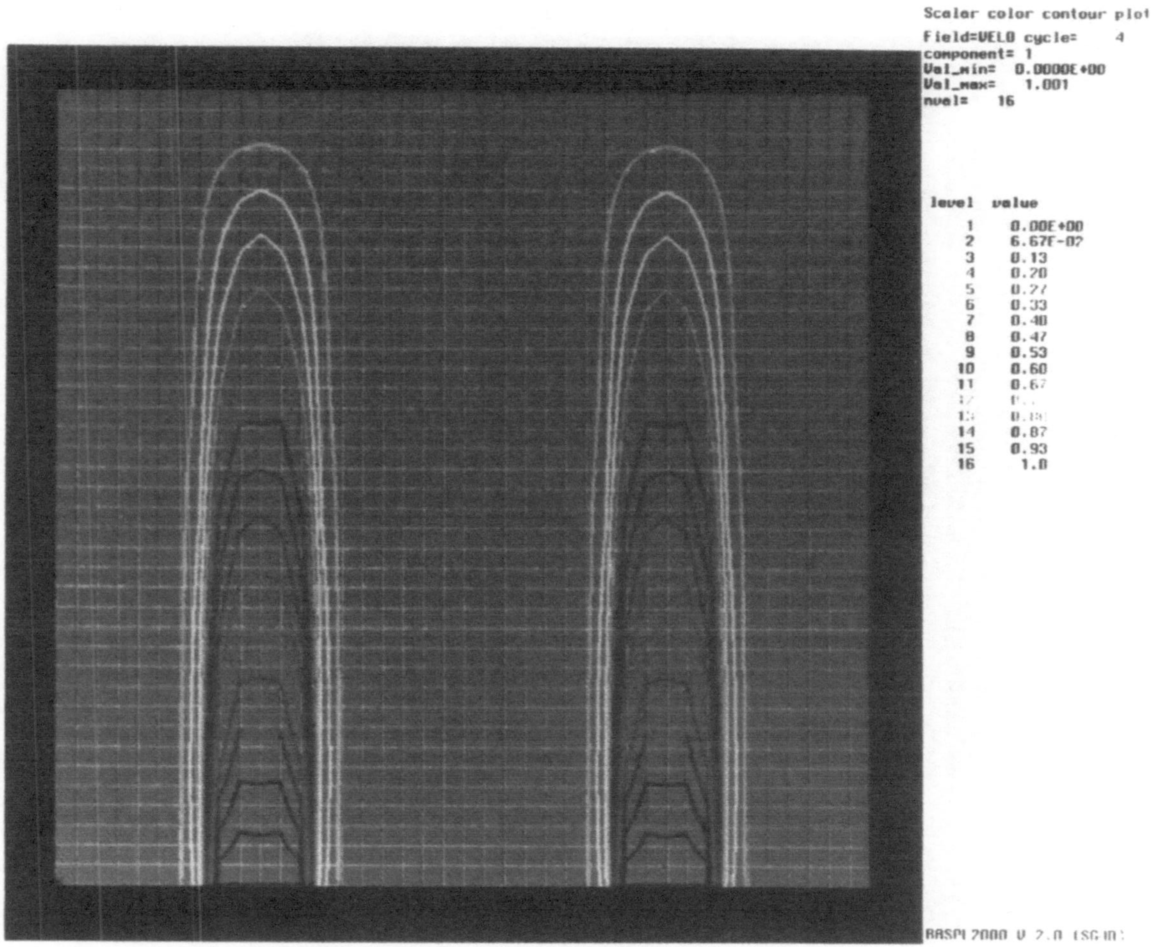

Scalar color contour plot

Field=VELO cycle=   4
component= 1
Val_min=   0.0000E+00
Val_max=   1.001
nvel=   16

level	value
1	0.00E+00
2	6.67E-02
3	0.13
4	0.20
5	0.27
6	0.33
7	0.40
8	0.47
9	0.53
10	0.60
11	0.67
12	
13	
14	0.87
15	0.93
16	1.0

Figure 10b

79

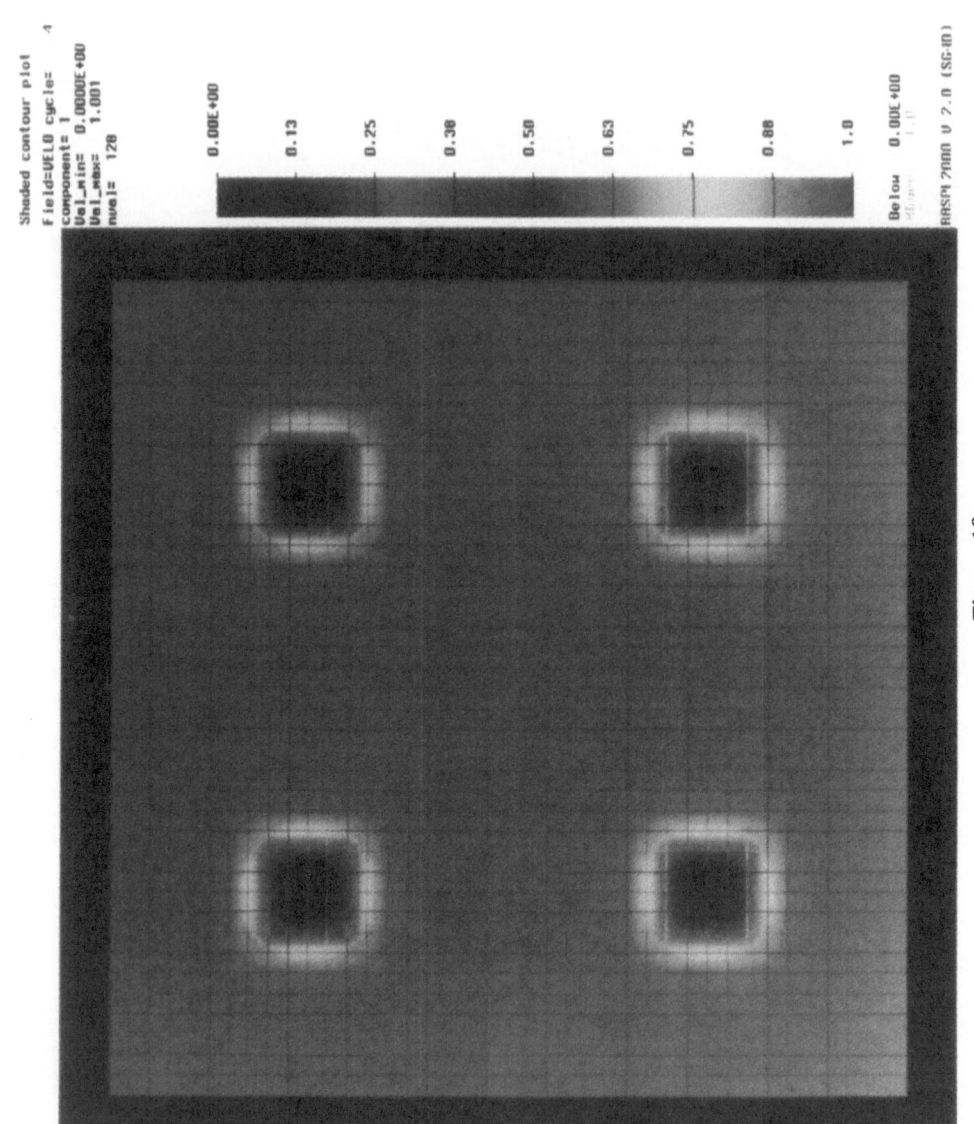

Figure 10c

9.6. BASPL

```
! problem of electric plate: BASPL script
!
db=elect.db
!
! plot the grid of elements
!
sele
  whole
end
sol= g
set squee [1 1 10]
plot
!
! plot a shaded contour plot of the potential
!
sele
  grid i=20      .
end
sol = cnf
self pot cyl 1
set polar [90 90 0]
set level co 128
plot
```

ACKNOWLEDGEMENTS

We are grateful to Dr. Tannenberger and Dr. Bossel for their guidance in the area of SOFC modellisation.

This work was supported by the Swiss National Science Foundation.

REFERENCES

1. R. Gruber et al, "Structured finite element and finite volume programs adapted to parallel vector computers", *Computer Physics Reports* 11:81-116 (1989).
2. M. Flück and R. Herbin, "Implementation of a parallel domain decomposition method for structured finite element problems", in *Proceedings of the Fifth International Symposium on Numerical Methods in Engineering, Lausanne, Switzerland*, R. Gruber, J. Periaux and R.P. Shaw, eds., Springer-Verlag (1989), Vol. 2, 299-304.
3. E. Bonomi, "Adaptive meshings of surfaces by structured grids using molecular dynamics", in *Proceedings of the Fifth International Symposium on Numerical Methods in Engineering, Lausanne, Switzerland*, R. Gruber, J. Periaux and R.P. Shaw, eds., Springer-Verlag (1989), Vol. 1, 113-120.
4. M. Flück and R. Herbin, "The ASTRID Finite Element Kernel", GASOV report 13, September 1988.
5. D.V. Anderson, W.A. Cooper, R. Gruber, S. Merazzi and U. Schwenn, "TERPSICHORE: A three-dimensional ideal magnetohydrodynamic stability program", these Proceedings.
6. S. Merazzi, "The MEMCOM user manual (Version 5.6)", SMR Corporation, Bienne, Switzerland (1988).

7. S. Merazzi, "The ASTRID data structure and processor handbook", GASOV Report 12, EPFL, Switzerland (1988).

8. S. Merazzi, "The BASPL postprocessing system", GASOV Report 14, EPFL, Switzerland (1989).

9. S. Merazzi, "The ASTRID System: an example of an integrated numerical analysis system", in *Proceedings of the 8th European Summer School in Computational Physics, Skalsky Dvur, CSSR, Sept. 1989*, to be published in Comput. Phys. Commun.

10. J.J. Desbiolles, J.J. Droux, A.F.A. Hoadley, J. Rappaz and M. Rappaz, "Modelling of solidification processes", in *Proceedings of the Fifth International Symposium on Numerical Methods in Engineering, Lausanne, Switzerland*, R. Gruber, J. Periaux and R.P. Shaw, eds., Springer-Verlag (1989), Vol. 1, 47-59 and 279.

11. S. Merazzi and P. Leyland, "A note on workstations and supercomputers", in *Supercomputing Review*, No. 1, Ecole Polytechnique Fédérale de Lausanne, Switzerland (1989).

12. U.G. Bossel and J.R. Ferguson, Natural Gas Fuelled Solid Oxide Fuel Cells and Systems, "Facts and Figures", Office Fédéral de l'Energie, July 1989.

13. M. Flück and R. Herbin, "The Mathematical Modelling of NG-fuelled SOFC, in Natural Gas Fuelled Solid Oxide Fuel Cells and Systems", *Proceedings of the Workshop on Mathematical Modelling, Charmey, Switzerland, July 1989*, (1989), 229-240.

14. ASTRID Group, presented by R. Gruber, "A simulation software for MIMD machines", in *Proceedings of the Fifth International Symposium on Numerical Methods in Engineering, Lausanne, Switzerland*, R. Gruber, J. Periaux and R.P. Shaw, eds., Springer-Verlag (1989), Vol. 1, 197-201

15. G. Meurant, "Iterative methods for multiprocessor vector computers", *Computer Physics Reports* 11: 51-80 (1989).

LARGE SCALE COMPUTATIONS IN SOLID STATE PHYSICS

P.E. Van Camp, V.E. Van Doren and J.T. Devreese[*]

University of Antwerp (RUCA), Groenenborgerlaan 171, B-2020
Antwerpen, Belgium

I. INTRODUCTION

In solid state physics calculations are called "ab-initio" or "first-principles" if they use as input only such data as the atomic number and weights of the elements making up the crystal, and fundamental physical constants such as the mass and charge of the electron and Planck's constant. A decade ago such calculations were impossible due to the computer limitations at that time. Nowadays such calculations have become a basic tool in solid state physics. At this time these calculations are restricted to mechanical, electronic and dielectric properties of crystals and crystalline surfaces at zero temperature.

In the present work the ab-initio non-local pseudopotential method with a plane wave basis is applied to cubic semiconductors. The basic quantity which is calculated is the equation of state, i.e. the relation between the crystalline volume and the total energy of the system. From this relation one can calculate the lattice constant, the bulk modulus and its pressure derivative. These quantities then determine the electronic properties such as the charge distribution, the band gap and its pressure derivative.

II. NUMERICAL PROCEDURES

The theoretical framework of the present work is the density functional theory in the local density approximation (see Appendix A). Using the pseudopotential theory with a plane wave basis (see Appendix B), one finally obtains the eigenvalue problem

$$HC = EC \tag{1}$$

where H is the matrix representation of the Kohn-Sham operator and the matrix C contains the wave function expansion coefficients in its columns. The diagonal matrix E contains the one-particle energies E_{kn}. It should be noted that Eq. (1) has to be solved iteratively since H depends upon ρ (see Eq. A7), which is calculated from the wave function expansion coefficients C (see Eq. B8).

---

[*] Also at: University of Antwerp (UIA), Universiteitsplein 1, B-2610 Antwerpen-Wilrijk and Eindhoven University of Technology, NL-5600 MB Eindhoven, The Netherlands.

Scientific Computing on Supercomputers II
Edited by J. T. Devreese and P. E. Van Camp
Plenum Press, New York, 1990

The numerical procedures which take up the bulk of the total run time are:
- the diagonalization of the Kohn-Sham matrix H for a given density ρ.
- the iterative solution of the self-consistent equation Eq. (1).

1. Matrix Diagonalization

In the present case the matrix H is complex Hermitian. This matrix is not diagonally dominant nor sparse. In the expansion of the electronic energy one only needs the valence band energies E_{kn} and the valence charge density ρ (see Eq. B9). Therefore we want a small number of the lowest eigenvalues and -vectors of a large matrix. For a typical material like GaAs we need the lowest four eigenvalues and -vectors of a 800×800 matrix.

The standard solution to this matrix problem is [11]:
- reduction to a real symmetric tridiagonal matrix (the Householder transformations)
- bisection method to obtain the lowest eigenvalues
- inverse iteration to find the corresponding eigenvectors followed by backtransformation.

The algorithm has been vectorized partly [12]. In Figure 1 we show timings of the standard solution as obtained on a two-pipe Cyber-205 and on an ETA-10P. A dramatic rise of the computer time is observed.

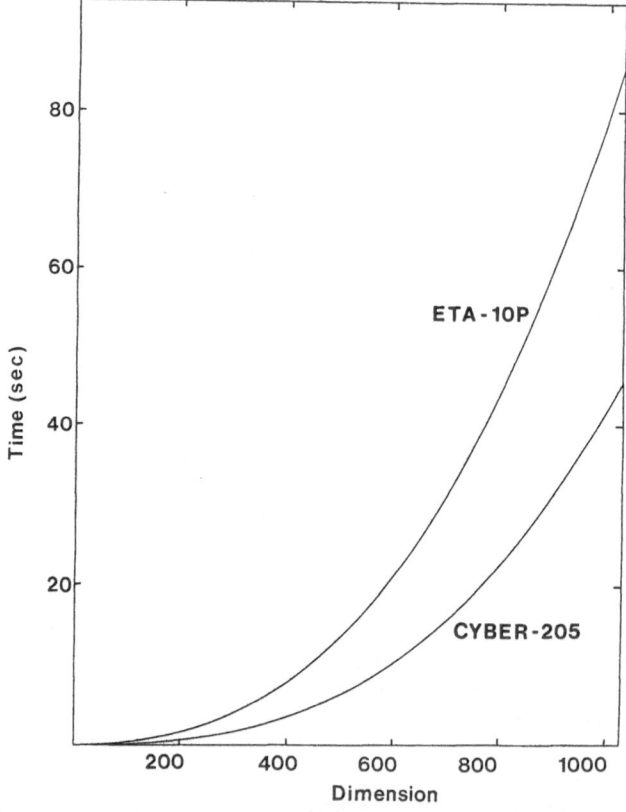

Figure 1. Time (in sec) of matrix diagonalization using the Householder scheme versus the dimension on the ETA-10P and on the Cyber-205.

For large dimensions there are several more efficient methods, all iterative in nature:
- a recursive method [13-15]
- the residual minimization-direct inversion in iterative subspace method (RMS-DIIS) [16]

1.1. The recursive method

In this method one constructs an orthogonal basis and recursion coefficients an and b_n from the relations (for n = 0, 1, ...):

$$b_{n+1} |n+1> = H|n> - a_n |n> - b_n |n-1>$$ (2)

where b_n is the normalization factor of $|n>$, $a_n = <n|H|n>$ and $b_0 = 0$. It should be noted that this procedure is nothing but the well-known Lanczos method [17] (expansion in the set $|0>$, $H|0>$, $H^2|0>$, ...) followed by a Gram-Schmidt orthonormalization. One can show that [14]

$$<n|H|n> = a_n \qquad \text{if } m = n$$

$$= b_n \qquad \text{if } m = n-1$$

$$= 0 \qquad \text{otherwise}$$ (3)

so that in this new basis the Kohn-Sham matrix has the simple tridiagonal form:

$$
\begin{pmatrix}
a_0 & b_1 & 0 & 0 & . & . \\
b_1 & a_1 & b_2 & 0 & . & . \\
0 & b_2 & a_2 & b_3 & . & . \\
0 & 0 & b_3 & a_3 & . & . \\
. & . & . & . & . & . \\
. & . & . & . & . & .
\end{pmatrix}
$$ (4)

the eigenvalues of which are the eigenvalues E_i of H in the subspace scanned by the states $|0> |1> ... |n>$. The corresponding eigenvector of H is then given by

$$\sum_{n=0} P_n (E_i) |n>$$ (5)

with

$$b_{n+1} P_n = (E - a_n) P_n - b_n P_{n-1}$$ (6)

and

$$P_{-1}(E) = 0; \quad P_0(E) = 1$$

It should be noted that these eigenvectors still should be orthogonalized. For the start vector |0> one can take an eigenvector of a truncated H-matrix. However, one must realize that, due to the fact that the expansion set is constructed by successive multiplication with H, the symmetry of all |n> is the same as |0>.

The whole procedure can be vectorized to a high degree since it mainly contains matrix-vector multiplications and dot products. Unfortunately the convergence is rather slow especially for low symmetry states. Furthermore the number of iterations grows rapidly for higher states. The use of a bigger start vector |0> reduces the number of iterations at the expense of more memory and of a longer start-up time (i.e. for the diagonalization of the truncated matrix).

1.2. The RMS-DIIS method

This method also starts from a vector obtained from the diagonalization of a truncated matrix. The extent to which this approximate eigenvector x_i and eigenvector b_i fail to be exact is the residual vector R:

$$R = (H - \lambda_i) |x_i> \tag{7}$$

In the m-th iteration the RMS-DIIS method then constructs a vector increment δx, using the residual vector R^{m-1} and eigenvalue λ^{m-1} of the previous cycle:

$$|\delta x> = - \sum_{i=1}^{N_0}{}' \frac{<x_i^0|R^{m-1}>}{\lambda_i^0 - \lambda^{m-1}} |x_i^0> - \sum_{i=N_0+1}^{N} \frac{<\rho_i|R^{m-1}>}{H_{ii}-\lambda^{m-1}} \tag{8}$$

with N_0 and N the dimensions of the truncated and full matrices. In order to determine what linear combination of the expansion vectors minimizes the residual of the new approximate eigenvector one constructs the matrices P and S given by

$$P_{ij} = <(H - \lambda^{m-1}) \delta x^i | H - \lambda^{m-1}) \delta x^j> \tag{9}$$

and

$$S_{ij} = <\delta x^i|\delta x^j> \tag{10}$$

The lowest eigenvalue of the (m+1) × (m+1) generalized eigenvalue problem

$$P|\alpha> = \rho^2 Q|\alpha> \tag{11}$$

determines the new approximate eigenvector

$$|x^{new}> = \sum \alpha_i |\delta x^i> \tag{12}$$

and eigenvalue of H

$$\lambda^{new} = <x^{new}|H|x^{new}> \tag{13}$$

The new residual

$$|R^{new}> = (H - \lambda^{new})|x^{new}> \tag{14}$$

is used as the criterion for stopping the iteration.

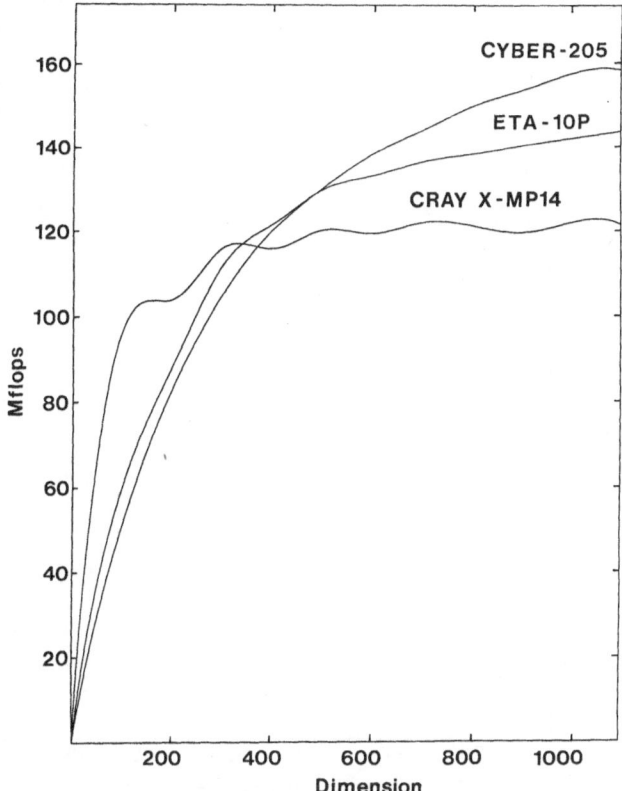

Figure 2. Number of Mflops for a matrix-vector multiplication versus the dimension on the Cyber-205, on the ETA-10P and on the CRAY X-MP14.

Tests on matrices of dimension ranging from 200 to 1000 have indicated that this method is very efficient: on the average 10 iterations are sufficient. The algorithm can be vectorized to a large extent due to the fact that most of the time is taken by matrix-vector multiplications and by dot products. Mflop ratings of matrix-vector multiplications on three supercomputers (Cyber-205, ETA-10P and the CRAY X-MP14) are shown in Figure 2.

The choice of the size of the truncated matrix is important since it influences the run time drastically. It is easily seen that the bigger the truncated matrix, the better the initial guess eigenvalue and -vector is but the longer it takes to diagonalize the small matrix. It also turns out that the number of iterations rises only modestly with decreasing small matrix size. A good rule of thumb is therefore to choose N_0 (the size of the truncated matrix) as small as possible but bigger than the number of eigenvalues wanted. One should also be careful to take it big enough so as to have the correct level ordering and degeneracies. Figure 3 shows timings of the diagonalization of a matrix with dimension 1040 versus the size of the small matrix.

A very important property of the RMS-DIIS method is that the computational time per iteration scales as N^2. The number of iterations is always much smaller than N so that this method is more efficient than the standard Householder procedure scaling as N^3. This is illustrated in Figure 4.

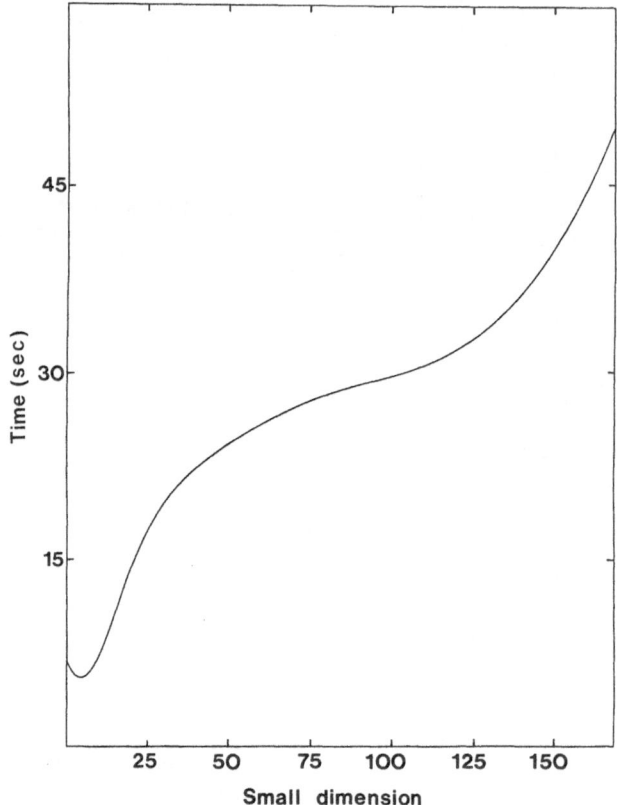

Figure 3. Time for the diagonalization of a 1040 × 1040 matrix versus the dimension of the small matrix.

2. Iterative Solution of the Self-Consistent Matrix

The total run time depends linearly on the number of iterations needed to solve the Kohn-Sham equations. As such it is therefore very important to find an efficient iteration scheme. Formally the equations to be solved can be written as

$$\rho = F[\rho] \qquad (15)$$

where ρ is a N-dimensional complex vector ($\rho \in C^N$) on F as a non-linear operator F: $C^N \to C^N$.

2.1. Simple iterations

Eq. (15) can be solved using

$$\rho^{n+1} = F[\rho^n] \qquad (16)$$

if for some n

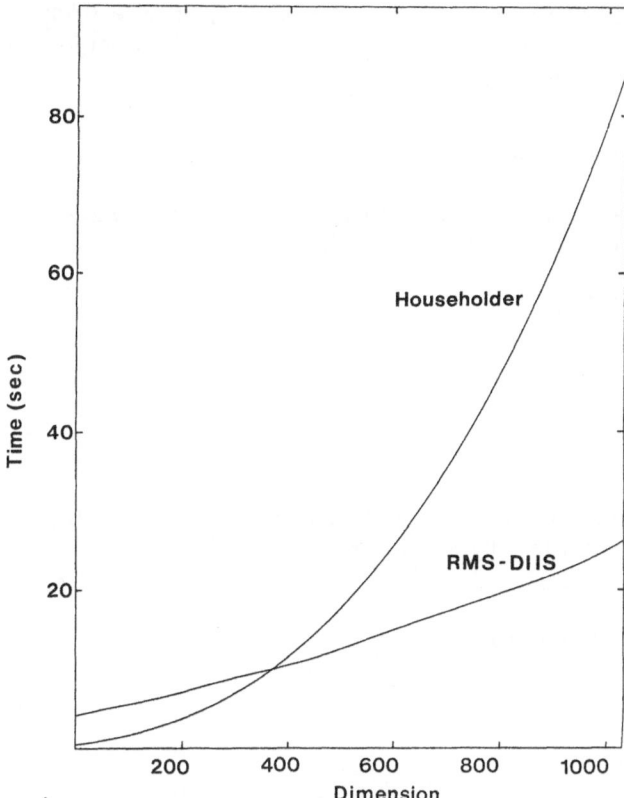

Figure 4. Comparison for the time for matrix diagonalization using the Householder and the RMS-DIIS methods.

$$||\rho^{n+1} - \rho^n|| < \varepsilon \qquad (17)$$

Unfortunately this is not always the case. Quite frequently the iteration Eq. (16) starts to diverge, while in most cases the convergence is too slow to be of practical value.

2.2. Mixing procedures

The simple mixing scheme often leads to an oscillating sequence of approximations. To damp these oscillations one can mix in one or more of the previous approximations leading to [18]:

$$\rho^{n+1} = \alpha_{n+1} F[\rho^n] + \alpha_n \rho^n + \alpha_{n-1} \rho^{n-1} + ... + \alpha_0 \rho^0 \qquad (18)$$

with

$$\sum \alpha_i = 1 \qquad (19)$$

If one mixes in only the previous approximant one has

$$\rho^{n+1} = \alpha F[\rho^n] + (1-\alpha) \rho^n \qquad (20)$$

In order to insure convergence one should take α small ($\alpha < 0.25$) resulting in a rather slow convergence. It is possible to optimize the rate of convergence of Eq. (20) or Eq. (18) leading to expressions for the unknown α [19].

2.3. An improved iteration scheme

An improved iteration scheme has been derived [20-21] based on the Taylor expansion $F[\rho]$ around the unknown solution ρ^*:

$$F[\rho] = \rho^* + J \, \delta\rho \tag{21}$$

with

$$J_{ij} = \frac{\partial F_i[\rho]}{\partial \rho_j} \tag{22}$$

Exact evaluation of J is hardly practical, although attempts have been published [22]. Let us therefore approximate J by a scalar λ_k such that

$$\sum_m |(\rho_m^{k+1} - \rho_m^k) - \lambda_k (\rho_m^k - \rho_m^{k-1})|^2 \tag{23}$$

is minimal, i.e.

$$\lambda_k = \frac{\sum \mathrm{Re}((\rho_m^{k+1} - \rho_m^k) - \lambda_k (\rho_m^k - \rho_m^{k-1})^*)}{\sum (\rho_m^k - \rho_m^{k+1}) - \lambda_k (\rho_m^k - \rho_m^{k-1})^*} \tag{24}$$

Eq. (21) then gives

$$\rho^* = (1 + \lambda_k) F[\rho] - \lambda_k \, \rho^n \tag{25}$$

Extrapolation then results in

$$\rho^* = \alpha F[\rho] - (1 - \alpha) \rho^n \tag{26}$$

with

$$\alpha = \frac{1}{1-\lambda} \tag{27}$$

We finally arrive at a scheme where mixing is done according to

$$\rho^{n+1} = \alpha \, \rho^{n+1} + (1 - \alpha) \rho^4 \tag{28}$$

and in the next step with

$$\rho^{n+2} = \alpha_{new} F[\rho^{n+1}] + (1 - \alpha_{new}) \rho^{n+1} \tag{29}$$

where

$$\alpha_{new} = \alpha \cdot \alpha_{old} \tag{30}$$

α being given by Eq. (27).

The scheme described above seems to work remarkably well. It converges faster than the mixing procedures and there is no need to guess α. It is easily vectorizable since it contains long (complex) vectors. In all cases where this scheme was used not a single divergency was observed.

III. SUMMARY OF THE RESULTS

In this section the main results of the ab-initio pseudopotential calculations on semiconductors will be summarized. Extensive discussions can be found in the literature on SiC and BN [23]; C, Si, Ge and α-Sn [24]; Ga-compounds [25]; BN [26-27]; In-compounds [28].

Table I gives the calculated and experimental lattice constants and bulk moduli for the IV-IV and III-V cubic semiconductors. It can be seen that the agreement with

Table I. Theoretical and experimental values of the lattice constant (in Å) and of the bulk modulus (in Mbar) of the cubic IV-IV and III-V semiconductors.

	a		B_0	
	Theory	Experiment	Theory	Experiment
C	3.5940	3.5668	4.08	4.42
Si	5.4276	5.4307	0.90	0.98
Ge	5.6535	5.6579	0.67	0.76
α-Sn	6.4483	6.4892	0.43	0.53
SiC	4.3925	4.360	1.96	2.24
BN	3.6246	3.615	3.98	3.69
BP	4.3573	4.538	1.62	1.65
BAs	4.7874	4.777	1.32	-
BSb	5.162	-	1.05	-
AlN	4.3791	-	1.76	-
AlP	5.4494	5.4635	0.85	0.86
AlAs	5.6536	5.660	0.71	0.78
AlSb	6.1208	6.1355	0.53	0.55
GaN	4.4415	-	1.65	-
GaP	5.3575	5.4505	0.87	0.89
GaAs	5.5769	5.6532	0.71	0.75
GaSb	6.0223	6.0959	0.53	0.55
InN	4.8669	-	1.31	-
InP	5.7107	5.8687	0.75	0.71
InAs	5.9156	6.0583	0.62	0.58
InSb	6.3599	6.4793	0.46	0.46

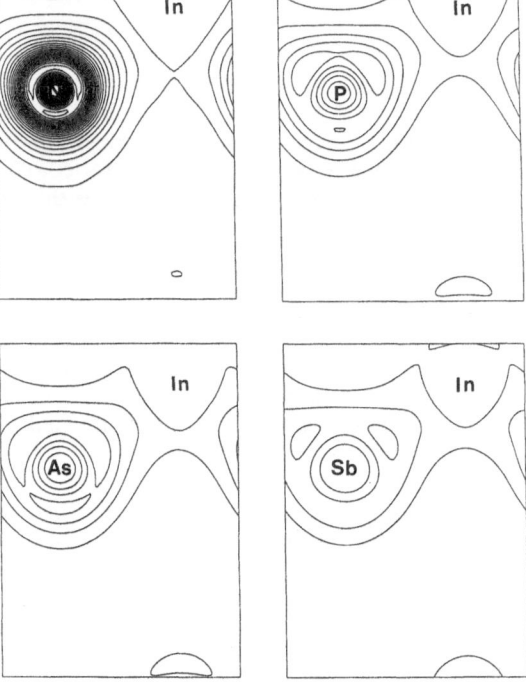

Figure 5. Valence charge densities in the [100]-plane of the cubic In-compounds (in e/au³).

experiment is very good. Slight deviations occur for the heavier compounds due to the influence of relativistic effects. It should be emphasized that the calculation is free of adjustable parameters.

Figure 5 displays the valence charge densities in the In-compounds. It can clearly be seen that InN is much more polar than InSb, i.e. the lighter the anion the more polar the crystal. The same trend is observed in the Al- and Ga-compounds.

In general one can say that a variety of equilibrium properties (at zero temperature) can be calculated with adequate accuracy. Calculated values which have not yet been measured can therefore be considered as trustworthy predictions. The increasing speed and capacity of supercomputers opens almost unlimited perspectives for this type of research.

IV. ACKNOWLEDGMENT

This work was performed in the framework of the "Institute for Materials Science" (I.M.S.) of the University of Antwerp (RUCA and UIA) funded by the IUAP 11 (Interuniversitaire Attractiepool 11 'Materials Science') of the Belgian Ministry of Scientific Affairs. ETA-10P and Cyber-205 supercomputer time was provided by the "NFWO - Supercomputer Project" of the Belgian National Science Foundation (Nationaal

Fonds voor Wetenschappelijk Onderzoek). We would like to thank the VUB-ULB for the opportunity to run benchmarks on their CRAY X-MP14.

APPENDIX A: THE DENSITY FUNCTIONAL THEORY

Density functional theory [1-4] is based on the Hohenberg-Kohn theorem which in its original form states that the ground-state energy of a system of spinless electrons is a unique functional of the electron density and that this functional has its minimum for the correct ground-state energy. This means that the ground-state energy of the electrons can be written as

$$E_e[n] = \int d^3r \ V_{en}(\bar{r},\bar{x}) \ \rho(\bar{r})$$

$$+ \frac{1}{2} \int d^3r \int d^3r' \ V_{ee}(\bar{r}-\bar{r}') \ \rho(r) \ \rho(r') + T[\rho] + E_{xc}[\rho] \qquad (A1)$$

where the first term is the potential energy of the electrons in the crystal potential with $\bar{x} = \{\bar{x}_{la}\}$, the second is the electrostatic energy, the third is the kinetic energy $T[\rho]$ and the last term is the exchange-correlation energy $E_{xc}[\rho]$, which are both unknown functionals of the density n. This density in the ground state is then found by minimizing E_e under the constraint that the total number of electrons remains constant. In principle, Eq. (A1) can be solved for any ground state configuration $\bar{x}$.

The additional assumption usually made is the local density approximation (LDA):

$$E_{xc}[\rho] = \int d^3r \ \rho(\bar{r}) \ \varepsilon_{xc}(\rho(\bar{r})) \qquad (A2)$$

For the exchange-correlation density ε_{xc} or the Kohn-Sham approximation:

$$\varepsilon_{xc} = \varepsilon_x + \varepsilon_c \qquad (A3)$$

with for the exchange

$$\varepsilon_x(\rho) = - \frac{3}{4} \left(\frac{3\rho}{\pi} \right)^{1/3} \qquad (A4)$$

and for the correlation [5]

$$\varepsilon_c(\rho) = - \frac{0.056\rho^{1/3}}{0.079 + \rho^{1/3}} \qquad (A5)$$

have been used throughout this work.

With any of these expressions into the exchange correlation functional (Eq. (A2)) the total electron energy E_e can be minimized leading to the one-particle Kohn-Sham equations:

$$H \ \psi_i(\bar{r}) = E_i \ \psi_i(\bar{r}) \qquad (A6)$$

with

$$H = -\frac{1}{2} \nabla^2 + V_{en}(\bar{r}) + \int d^3r' \, V_{ee}(\bar{r}-\bar{r}') \, \rho(\bar{r}) + \frac{\delta E_{xc}[\rho]}{\delta \rho} \qquad (A7)$$

APPENDIX B: THE PSEUDOPOTENTIAL THEORY AND PLANE WAVE EXPANSION

The density functional theory is usually combined with pseudopotential theory in which the core electrons (i.e. the electrons occupying the completely filled shells of the atoms) are taken together with the nuclei to make up for V_{en}. Therefore in an atom the effective potential experienced by a valence electron is due on the one hand to the nucleus together with the core-electrons and on the other hand to the remaining valence electrons. In a solid these core-electrons remain very localized around the nucleus whereas the outer valence electrons determine largely the properties of the solid. Most crystal pseudopotentials are constructed from atomic calculations and these are available in the literature. As it turns out a serious deficiency of these pseudopotentials is that the total electron density outside the core region is too low; however this charge density is an important ingredient of the Kohn-Sham equations (Eq. (A6)).

Recently, this problem has been solved by the construction of ab-initio self-consistent pseudopotentials for the atoms [6]. The pseudopotential is constructed in such a way as to reproduce correctly the electron energies and wavefunctions of the valence electrons and therefore also the valence charge densities outside a chosen core region. The correct reference values are obtained from relativistic all-electron calculations. These pseudopotentials are angular-momentum (or energy) dependent, i.e. non-local. Furthermore they yield the correct amount of charge inside the core region so that their ionic electrostatic potential outside the core has also the correct value. This property is called "norm-conservation". Also, deviations of the scattering amplitudes for energies in the neighborhood of the value used for the construction of the pseudopotential from the exact values are small. All these properties make the ab-initio norm-conserving non-local atomic pseudopotentials transferable i.e. they can be used in any configuration of condensed matter provided the core regions of neighboring atoms do not overlap. In the present calculations, the ionic ab-initio norm-conserving pseudopotential of Bachelet, Hamann and Schlüter (BHS) [6] is used:

$$V_{en}(\bar{r}) = V_{core}(\bar{r}) + \sum_{l} \Delta V_{l}(\bar{r}) \, P_{l} \qquad (B1)$$

where P_l is the projection operator for angular momentum l.

As explained in Ref. 6 the numerical values of this potential are fitted, with high precision, to the following expression:

$$V_{core}(\bar{r}) = -\frac{Z}{r} \, \{ \sum_{i=1}^{2} c_i^{core} \, erf[(\alpha_i^{core})^{1/2} \, r] \} \qquad (B2)$$

and

$$V_{l} = \sum_{i=1}^{3} (A_i^l + r^2 \, A_{i+3}^l) \, e^{-\alpha_i^l r^2} \qquad (B3)$$

The coefficients A_i^l and exponents α_i are tabulated in Ref. 6. However, the calculation of the coefficients A_i^l requires special care [7].

For crystalline systems the electron wavefunctions are expanded in plane waves:

$$\psi_{n,\bar{k}}(\bar{r}) = \sum_{\bar{G}} C_{n,\bar{k}}(\bar{G}) \, e^{i(\bar{k}+\bar{G})\bar{r}} \tag{B4}$$

where n is the electron band index, $\bar{k}$ the wave vector in the Brillouin zone and $\{\bar{G}\}$ the set of reciprocal lattice vectors generated by the space group appropriate for the crystal. In this basis the Kohn-Sham equations take on the following matrix form:

$$H_{\bar{G}\bar{G}'}(\bar{k}) = \frac{1}{2} (\bar{k}+\bar{G})^2 \, \delta_{\bar{G}\bar{G}'} \, \frac{4\pi}{|\bar{G}-\bar{G}'|^2} \rho(\bar{G}-\bar{G}')$$

$$+ V_i(\bar{G},\bar{G}')S(\bar{G}-\bar{G}') + \int d^3r \, e^{i(\bar{G}-\bar{G}').\bar{r}} \frac{E_{xc}[\rho]}{\delta\rho} \tag{B5}$$

where

$$S(\bar{G}-\bar{G}') = \sum_a e^{i(\bar{G}-\bar{G}').R_a} \tag{B6}$$

$$V_i(\bar{G},\bar{G}') = \int d^3r \, e^{-i\bar{G}r} \{V_{core}(\bar{r}) + \sum_1 \Delta V_1(\bar{r}) \, P_1\} \, e^{i\bar{G}'\bar{r}} \tag{B7}$$

and

$$\rho(\bar{G}) = \sum_{nk\bar{G}'} C_{n\bar{k}}(\bar{G}') \, C_{n\bar{k}}(\bar{G}+\bar{G}') \tag{B8}$$

P_1 is a projection operator for angular momentum 1. Then, the total electronic energy can be expressed in terms of the one-particle Kohn-Sham energies as follows

$$E_e[n] = \sum_{val} E_{n\bar{k}} - \frac{1}{2} \int d^3r \int d^3r' \, \rho(\bar{r}) \frac{1}{|\bar{r}-\bar{r}'|} \rho(\bar{r}')$$

$$+ \int d^3r \, \rho(\bar{r}) \{\varepsilon_{xc}(\bar{r}) - \frac{\delta E_{xc}[\rho]}{\delta\rho}\} \tag{B9}$$

The total energy of the system E_T is obtained by adding the pure electrostatic energy of the cores to the electronic energy (Eq. (B9)).

The equation of state of the crystal

$$p = -\frac{dE_T}{dV} \tag{B10}$$

can then be determined from the calculated total energies for different values of the volume V. In order to avoid numerical differentiation, either the Murnaghan equation of state [8]

$$p = \frac{B_0}{3} \left[\left(\frac{V_0}{V} \right)^{B_0'} - 1 \right],$$ (B11)

the Birch equation of state [9]

$$p = \frac{2}{3} B_0 \left[\left(\frac{V_0}{V} \right)^{7/3} - \left(\frac{V_0}{V} \right)^{5/3} \right] \left\{ 1 + \frac{3}{4} (B_0' - 4) \left[\left(\frac{V_0}{V} \right)^{2/3} - 1 \right] \right\}$$ (B12)

or other empirical equations (e.g. the Vinet equation [10]) are fitted to the calculated points B_0 and B_0' are the bulk modulus and its pressure derivative and V_0 is the volume at zero pressure.

REFERENCES

1. See e.g. in *Electronic Structure, Dynamics and Quantum Structural Properties of Condensed Matter*, J.T. Devreese and P.E. Van Camp, eds., Plenum Press, New York (1985).
2. P. Hohenberg and W. Kohn, *Phys. Rev.* 136:B864 (1964).
3. W. Kohn and L.J. Sham, *Phys. Rev.* 140:A1133 (1965).
4. L. Sham and W. Kohn, *Phys. Rev.* 145:B561 (1966).
5. E. Wigner, *Phys. Rev.* 46:1002 (1934).
6. G. Bachelet, D. Hamann and M. Schlüter, *Phys. Rev.* B26:4199 (1982).
7. P.J. Denteneer, W. Van Haeringen, F. Brosens, J.T. Devreese, P.E. Van Camp, V.E. Van Doren and O.H. Nielsen, *Phys. Rev.* B37:4795 (1988).
8. F. Murnaghan, *Proc. Nat. Acad. Sci.* USA 3:244 (1944).
9. F. Birch, *Geophys. Res.* 57:227 (1952).
10. P. Vinet, J. Ferrante, J. Smith and J. Rose, *J. Phys.* C19:L467 (1986).
11. B. Garbow, J. Boyle, J. Dongarra and C. Moler, *Lecture Notes in Computer Science*, volume 51 (1977).
12. P.E. Van Camp and J.T. Devreese, in *Scientific Computing on Supercomputers*, J.T. Devreese and P.E. Van Camp, eds., Plenum Press, New York (1989).
13. R. Haydock, V. Heine and M. Kelly, *J. Phys.* C5:2845 (1972).
14. V. Heine, *Solid State Physics* 35:87 (1980).
15. U. Brüstel and K. Unger, *Phys. Stat. Sol.* (b) 123:229 (1984).
16. D. Wood and A. Zunger, *J. Phys.* A18:1343 (1985).
17. C. Lanczos, *J. Res. Nat. Bur. Stand.* 45:255 (1950).
18. See e.g. P. Dederichs and R. Zeller, *Phys. Rev.* B28:5462 (1983).
19. H. Akai and P. Dederichs, *J. Phys.* C18:2455 (1984).
20. M.J. Dewar and P.K. Weiner, *Comp. Chem.* 2:31 (1978).
21. F. Badziag and F. Solms, *Comp. Chem.* 12:233 (1988).
22. See e.g. G. Srivastava, *J. Phys.* A17:L317 (1984).
23. P.E. Van Camp, V.E. Van Doren and J.T. Devreese, *Phys. Stat. Sol.* (b) 146:573 (1988).
24. P.E. Van Camp, V.E. Van Doren and J.T. Devreese, *Phys. Rev.* B38:12675 (1988).
25. P.E. Van Camp, V.E. Van Doren and J.T. Devreese, *Phys. Rev.* B38:9906 (1988).
26. P.E. Van Camp, V.E. Van Doren and J.T. Devreese, *Solid State Commun.* 71:1055 (1988).

27. P.E. Van Camp, V.E. Van Doren and J.T. Devreese, in *Proceedings of the 12th AIRAPT - 27th EHPRG Conference, Paderborn* (1989).

28. P.E. Van Camp, V.E. Van Doren and J.T. Devreese, *Phys. Rev.* B41:432 (1990).

COULD USER-FRIENDLY SUPERCOMPUTERS BE DESIGNED?

Willi Schönauer and Reinhard Strebler

Rechenzentrum der Universität Karlsruhe, Postfach 6980, D-7500 Karlsruhe 1, F.R.G.

ABSTRACT

The requirements for a supercomputer in a versatile engineering scientific computing environment are specified. Then the present trend to an "external" parallelism is discussed: MIMD architectures of parallel processors shift all problems to the users. But high performance needs parallelism. In the form of the propositon for a Continuous Pipe Vector Computer (CPVC) the ideas for a user-friendly SIMD multipipeline architecture are presented. The special design minimizes the number of lost cycles and thus delivers for sufficiently large problems a sustained performance near to the peak performance of the hardware technology. The purpose of this paper is to demonstrate that and how parallelism can be shifted to the hardware level and thus is transparent to the user.

1. INTRODUCTION

We want to discuss scientific supercomputing in an industrial R & D environment or in the environment of a technical university. Such an environment is characterized by a versatile job profile. Most of the problems result from discretization methods for the solution of systems of nonlinear PDEs (partial differential equations). Consequently most of the computation is used for the direct or iterative solution of full or sparse large linear systems of equations, for matrix-vector or matrix-matrix operations on full or sparse matrices of all kinds. Supercomputers that are well suited to treat such types of problems are suited to deal as well with problems of other sciences, e.g. theoretical physics and chemistry. This is the market of supercomputers and the background for our architectural discussion. It is absolutely necessary to define clearly the purpose a supercomputer is aimed at.

The selection and assessment of a supercomputer in such a scientific and engineering computation environment is made by a benchmark that is composed from typical codes of the most important users. This again reflects the versatile job profile. Such a benchmark then measures the <u>throughput</u> of a supercomputer, i.e. the efficiency for the test job profile and thus for the large investment of <u>existing</u> software. Besides the untuned benchmark the manufacturer may run a "tuned" benchmark that has been adapted to the special architecture of the supercomputer. So you may see what you

Scientific Computing on Supercomputers II
Edited by J. T. Devreese and P. E. Van Camp
Plenum Press, New York, 1990

can gain in performance if you tune your software correspondingly. But this tells you also that you <u>must</u> do something if you want to use your supercomputer with high efficiency. The "quality" of a super-computer then can be measured by the amount of work that must be done to obtain an acceptable use of the tested supercomputer. Think, e.g., that you put your industrial benchmark on a massively parallel computer. Then you will be far from the theoretical peak performance and you must carefully redesign your <u>whole</u> software if you want to get close to the expected performance. This situation illustrates what we mean by "user-friendly". In this paper we want to discuss supercomputing entirely from the point of view of a <u>user</u>.

We shall briefly discuss the requirements of the engineers and the basic types of parallelism, then the ideas of the Continuous Pipe Vector Computer (CPVC) will be developed, followed by some concluding remarks. The ideas of the CPVC have been presented for the first time as section 18 in [1], an extended version is [2,3]. The present paper develops the basic ideas further and discusses some problems in more detail.

2. THE REQUIREMENTS FOR A SUPERCOMPUTER IN ENGINEERING SCIENCES

2.1. Performance and Balanced System

We have gained much experience with the solution of PDEs with different discretization methods on different vector computers. We need for the solution of a system of 6 PDEs on a three-dimensional grid of 50 x 50 x 50 grid lines by a computer with 100 MFLOPS (megaflops = million floating point operations per second) <u>sustained</u> rate roughly 1 h CPU-time (the reasonable limit) and we need 64 Mwords (million words) of 64 bits main memory to store the operands in order to keep the arithmetic pipes busy (waiting for disks is not possible). For more details of this "Typical Example" see [1]. The fundamental relation that we obtain from this example for discretization problems is the relation:

100 MFLOPS <u>sustained</u> need 64 Mwords main memory.

This is what we call a balanced system.

But engineers need for sufficient details grids of 500 x 500 x 500 grid lines. Because the type of linear system has a band-structure so that the storage and computation increase linearly with the number of unknowns we need 1000-fold resources, which means, scaling the above fundamental relation,

100 GFLOPS <u>sustained</u> need 64 Gwords main memory.

Here GFLOPS = gigaflops = 1000 MFLOPS, Gwords = 1000 Mwords. The crucial point in this relation are the 64 Gwords. They will determine the price of the supercomputer. Only if memories of 64 Gwords are available and/or can be afforded, the computing power of the 100 GFLOPS can be used for such types of engineering problems. There might be special problems that need less memory for 100 GFLOPS, but such a computer would not be a balanced system for engineering problems and thus may have a narrow market. On the other hand, a computer with sufficient memory can handle easily problems with less memory requirements.

2.2. Data Transfer Operations

The development of the FIDISOL (Finite Difference Solver, see section 17 in [1]) and of the VECFEM (Vectorized Finite Element [4,5]) program packages has demonstrated the importance of the data structures for an efficient vectorization. The design of a vector computer program must start with the choice of the data structures. But different tasks in a large program may need different data structures. This leads us to the following basic rule: We have to separate the selection of the data from the processing of the data, i.e. to establish at first the optimal data structure for each subtask, and then to process. This shows the absolute necessity of efficient hardware for data transfer operations, namely mask-controlled operations of the type pack, unpack, merge, and indexvector-controlled operations gather/scatter (indirect addressing). Only a vector computer with exellent data transfer operations has a chance to process efficiently data for sparse matrices or unstructured grids. The early CRAY computers had no data transfer operations, so corresponding operations were executed in scalar mode. The recent CRAY models still have a poor set of data transfer operations. The defunct CYBER 205 and ETA 10 computers had the richest set of data transfer operations.

2.3. Scalar Performance

There is always a remainder of scalar operations that can be executed only by the scalar unit(s). The speed of the scalar unit can be increased only by reducing the cycle time. For 100 nsec technology we ·need for scalar operations 25 times the execution time of 4 nsec technology. Thus we need the fastest cycle time for a fast supercomputer.

2.4. Programming Language

In an industrial R & D environment there are large investments in software and in know-how and there is the use of large commercial program packages, and in a university environment there is the use of the large international scientific program libraries and there is the exchange of programs with other universities. The common basis of all these applications is Fortran in whatever form: Fortran 77, Fortran 8x, 9x, ... We will never get rid of the Fortran history in scientific computing. Thus a supercomputer must be programmable in portable Fortran without special extensions that destroy the portability. All considerations concerning other languages in this environment are illusory. For most of the installations the accumulated software is much more expensive than the hardware. The hardware will periodically be replaced, the software will develop gradually.

2.5. Summary of Requirements

As mentioned above it is necessary to define clearly the purpose and environment of a supercomputer because these points determine the requirements. We discuss in this paper a supercomputer for scientific computation in an engineering environment. The properties that we expect are the following, summarizing the preceding discussion:

- Sustained performance of 100 GFLOPS and main memory of 64 Gwords (a word in this environment has 64 bits).

- Excellent data transfer operations of type pack, unpack, merge, gather, scatter.

- Fastest technology with fastest cycle time.

- Programmable in portable Fortran without special extensions.

We now may ask: How should a computer with these properties look like so that it has a user-friendly architecture?

3. PARALLEL ARCHITECTURES

If we want a performance of 100 GFLOPS we need parallel pipelines. The CRAY-2 has 4.1 nsec cycle time and thus a parallel execution of addition and multiplication (supervector speed) in a <u>single</u> pipe group delivers a theoretical peak performance of 0.488 GFLOPS, the CRAY-3 with 2 nsec is expected for 1990 and delivers 1 GFLOPS and the CRAY-4, expected in 1994/95, with 1 nsec would deliver 2 GFLOPS. This is for <u>one</u> "combination pipe", i.e. a group of an addition and a multiplication pipe (pipe group). Therefore we need parallel pipelines or processors for 100 GFLOPS. The question is now, how this parallelism should be organized. For this purpose we briefly discuss basic architectural concepts (see [6]).

W. Giloi, the hardware architect of the SUPRENUM parallel computer, states that parallel micro-processors are not competitive, one has to use parallel pipes [8]. This means that the innermost hardware kernel executes vector operations and thus the innermost loop must be vectorizable for an efficient use of such a parallel computer. In Figure 3.1 the three basic architectures for parallel computers are depicted, these are all MIMD (multiple instruction stream, multiple data stream) architectures. We now discuss briefly and in a very simplifying manner the properties of these basic parallel architectures. Note that these types of computers have all an "external" parallelism: The user has to adapt his program to the special architecture, the parallelism is not transparent to the user.

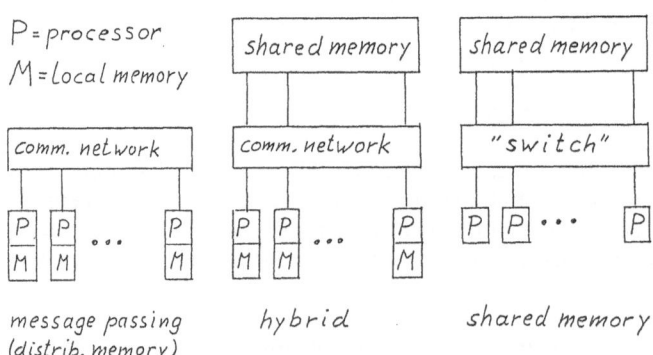

Figure 3.1. The three basic architectures for MIMD parallel computers.

<u>Message passing systems</u> can be characterized as follows:

> Message passing systems can easily be
> built, but they cannot be programmed.

Presently every year many conferences on parallel computers and above all on message passing systems take place and the literature is abundant in excellent papers on this subject, therefore we do not list here all the corresponding references. Researchers report high speedups $s = T_1/T_N$ (T_1 = time for one processor to solve a certain problem, T_N = time for N processors), but this means only that the system is an excellent <u>special purpose</u> computer. The real problem is the permanent usage of the processors for a <u>versatile</u> job profile. In message passing systems the computation <u>and</u> data is distributed onto the processors and (local) memories, there is no central or shared memory. An "easy" program switch to another program to use idle processors is not possible. Consequently message passing systems are "single user systems", i.e. PCs, or if they are large, they are GFLOPS-PCs. An efficient global resource management is <u>not</u> possible. This means "cannot be programmed". A subdivision into fixed partitions is in principle the same. SUPRENUM and Intel's iPSC/2 are of this type of architecture.

<u>Shared memory systems</u> can be characterized as follows:

> Shared memory systems can easily be
> programmed, but they cannot be built.

In shared memory systems only the computation must be distributed onto the processors. But the processors access the shared memory with pipeline consumption rate. For the uncoordinated access of MIMD this works only for a few processors. Thus the increase in performance is restricted by the memory bandwidth and the memory contention of the processors.
But here at least a central resource management is possible. Multiprocessor supercomputers like CRAY-2, CRAY X-MP, Y-MP fall under this category if they are used in a multitasking style.

<u>Hybrid systems</u> have local and shared memory. They combine the <u>disadvantages</u> of both systems, see Karp [9]. The CEDAR [10] is of this type.

Concluding this very brief and coarse discussion of parallel architectures, we might say: With MIMD all problems are shifted to the user. <u>He</u> has to distribute the computation (and the data) to the different processors, he is responsible for a good usage of the computer. MIMD introduces <u>time</u> as an essential factor: One has to know which data is where at what time [9]. Is this a user-friendly architecture?

4. THE CONTINUOUS PIPE VECTOR COMPUTER (CPVC)

In order to give a positive critism we present the ideas for a supercomputer with a <u>user-friendly</u> parallelism, so that the user sees only a SIMD (single instruction stream, multiple data stream) monoprocessor. The only purpose of this presentation is to demonstrate that user-friendly architectures <u>can</u> be developed. These ideas can also be used to "measure" actual designs and to demonstrate their deficiencies. If we want to obtain 100 GFLOPS sustained performance we need parallelism. The question is only, <u>how</u> to organize the parallelism in an optimal way. In the following we present the different considerations that finally form the CPVC.

4.1. Memory Bandwidth

The most important operation is the vector triad

$$d_i = a_i + b_i * c_i \tag{4.1}$$

that needs 3 loads and one store, i.e. 4 memory references per cycle and pipe. Usually one has an equidistribution of + and *. Then the addition and multiplication pipes can operate simultaneously, producing supervector speed, i.e. 2 results per cycle and "combination pipe" or pipe group. A special case is the so called linked triad with one scalar operand

$$c_i = a_i + s * b_i \tag{4.2}$$

that needs two loads and one store per cycle and pipe. A still more special case is the so called saxpy operation, the contracting linked triad (repeated)

$$b_i = b_i + s * a_i \tag{4.3}$$

that needs only one store if b_i is "fixed" in a vector register for a register-to-register machine.

Only if the bandwidth for the vector triad (4.1) is met, the computer has a real chance to come near to the theoretical peak performance. Then a flexible use of the addition and multiplication pipe for parallel operation (supervector speed) is possible on the Fortran level, i.e. without special "tricks" on the assembler language level. The CRAY-2, e.g., can be considered as a "saxpy machine" because it has the "one word per cycle and pipe bottleneck" between main memory and vector registers.

Our formula for a "realistic performance" of a <u>single</u> pipe (group) is

$$r_1 = \frac{1\ 000}{\tau\ [\text{nsec}]} * 2 * \frac{m}{4} * d * f\ [\text{MFLOPS}]. \tag{4.4}$$

where $1000/\tau[\text{nsec}]$ is the theoretical peak performance of a single pipe, measured in MFLOPS if the cycle time τ is in nsec, the factor "2" gives supervector speed for the addition/multiplication combination. $m/4$ is the influence of the memory bandwidth where m is the number of available memory references per cycle and pipe (group), d is the influence of the finite mean vector length, with $0 < d \le 1$, $d = 1$ for vectorlength $n = \infty$, and f cares for the mean "usage" of the pipe(s), with $0 < f \le 1$, f = 1 for permanent use, f is the part of the CPU-time in which the pipe(s) is (are) busy except the influence of the memory bandwidth. It is clear that d and f depend not only on the hardware but also on the type of problem to be solved and on its data structure. So we want a supercomputer with m = 4 memory references per cycle and pipe in order to make m/4 = 1 in (4.4). This bandwidth gives also the greatest flexibility in the choice of the algorithm. Up to now no supercomputer does exist with a memory bandwidth of 4 words per cycle and pipe group. The best value is 3 words per cycle and pipe group for the CRAY X-MP and Y-MP.

4.2. Local and Extended Memory

In large supercomputers the size of the memory has a significant influence on the price. Nobody in the near future can afford a main memory of 64 Gwords (with 64 bits) with the large bandwidth of 4 memory references per cycle and pipe (group). But the practical experience has told us that only about 10 % of the memory must be directly accessible with this bandwidth, the remaining 90 % can be accessed sequentially like a file, preferably in blocks. So we can subdivide the main memory into 10 % local memory with the high bandwidth and 90 % extended memory. The essential condition for the extended memory is, that it has a memory bandwidth of one word per cycle and pipe (group) to the local memory. Then one operand "in the mean" can be obtained via a buffer from the extended memory or alternatively (not simultaneously) one result can be delivered to the extended memory. Only under this condition the extended memory is a real "extension" of the main memory. The organization of the data in the extended memory should be made explicitly by the user, not automatically like for a virtual memory. For a fast computer the danger of thrashing is too serious. The SSD (Solidstate Storage Device) of CRAY or the expanded memory of the IBM 3090 are examples of an extended memory, but none has a sufficient bandwidth. So we can subdivide for our example of the CPVC the 64 Gwords main memory into 6.4 Gwords local and 57.6 Gwords extended memory.

4.3. Number of Pipes

For 100 GFLOPS sustained we need parallel pipes. If we assume a cycle time of 1.28 nsec that might be obtained in 1994 (?) and we take because of the special design of the CPVC the basis of supervector speed we need 64 pipe groups of addition/multiplication pipes. This is realistic only for an optimal architecture with optimal factors in the performance formula (4.4). The performance of a N-pipe (group) supercomputer is

$$r_N = N * r_1 \text{ [MFLOPS]}, \tag{4.5}$$

with r_1 from (4.4).

In (4.4) the factor d accounts for the influence of the finite vector length that changes for "conventional" parallel computers with the number of parallel pipes, see below. But for MIMD parallel computers, i.e. computers with "external" parallelism, the factor f in (4.4), that gives the fraction of time that the pipes are busy (except waiting due to a memory bottleneck), may change considerably with the number of pipes = processors. It may happen that an increase of N reduces f correspondingly so that no net gain remains. Thus the factors d and f in (4.4) depend strongly on N, i.e. a pipe operated as a single pipe may deliver a much higher performance than the same pipe in a certain multipipe/multiprocessor architecture. It is interesting to note that the CRAY-4 is expected in 1994/95 with 1 nsec cycle time and 64 pipes, giving a theoretical peak performance of 128 GFLOPS. But only what will remain as a sustained performance for non-saxpy problems is the realistic "benchmark-performance" of the CRAY-4 and this is the only performance the user is interested in.

We should recall that a 100 GFLOPS sustained supercomputer should have a 64 Gword main memory. We do not expect to have an "affordable" memory of such size before 1994. This corresponds also to the time scale when 1.28 nsec cycle time can be expected. In 1994 the 64 Mbit chip may be available. This means that on a single chip are 8 MB or 1 Mword. Nevertheless for 64 Gwords we need 64 000 of those 64 Mbit chips. This illustrates the meaning of 64 Gwords. But if such a memory is not included in the CPVC, the 100 GFLOPS cannot be used for engineering problems. This means also that a CPU speed of 100 GFLOPS, obtained by whatever type of architecture, e.g. massively parallel processors, is useless without such a memory size. So we stress again that memory is the real problem, not CPU speed.

Presently neither a cycle time of 1.28 nsec nor an "affordable" memory of 64 Gwords is available. Therefore presently a balanced 100 GFLOPS supercomputer yet cannot be built. In the meantime we should use slower technology that then clearly delivers correspondingly less GFLOPS. But such slower technology should nevertheless be used in order to access a broader market. These points are discussed in more detail below in section 4.6.

4.4. Memory Organization

Here we discuss the organization of the local memory that needs 4 data paths (3 loads, 1 store) with 64 words (with 64 bits) per cycle, each, to keep the 64 pipe groups busy. With an uncoordinated MIMD access this could not be built. Here we develop the ideas for a special <u>synchronous</u> access. It is clear that the parallelism of the N = 64 pipes includes a corresponding parallelism of busses. In Figure 4.1 we demonstrate the principle for N = 4 (instead of N = 64) pipes. We assume a memory of M = k * N memory banks and 4 bus groups with N busses, each. Each single bus can transfer 1 word per cycle, so we have 4 * N words per cycle for the N pipe groups. Each bus is connected to every Nth bank, each bank is connected to one bus of each group. Because the operands must arrive at the correct time and because of possible bank conflicts the number of banks should be large enough and the bank busy time (bbt) should be as small as possible. The minimum number of banks is 4 * 64 * bbt, if we count bbt in number of cycles. Realistic values of bbt are 5 to 10 cycles. Thus the minimum number of banks would be 1280 to 2560. But one should have roughly twice these numbers, thus 2560 to 5120 banks would be appropriate for 5 to 10 cycles bbt. In the latter case in one bank are 12.5 Mwords which is still fairly large. Thus 10240 banks would be preferable that not too much data is in a single bank. Nevertheless delay registers are needed for synchronization of the operands.

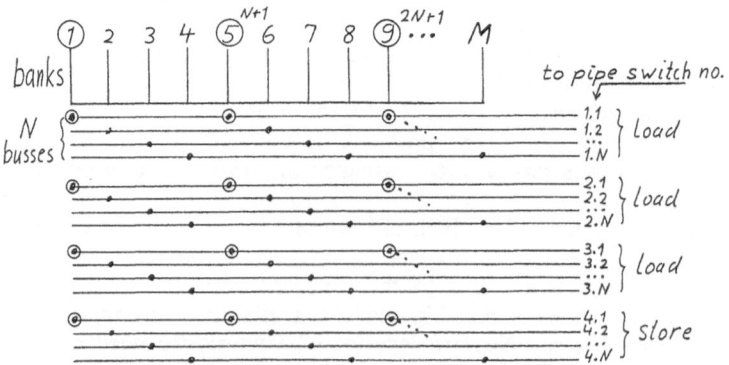

Figure 4.1. Connections between memory banks, busses and pipe switch.

Connected banks and busses may be locally grouped together on chips and boards. It should be mentioned that the local memory must also deliver a port for a blockwise transfer of 1 word per cycle and pipe to the extended memory. This port should have a lower priority than the busses to the pipe switch.

The memory structure of Figure 4.1 has a consequence on the meaning of the digits of an address. We explain this for a "decimal machine" (in reality we should discuss binary addresses). If we have the following address (x,y,z denote digit positions)

$$x \ x \ x \ x \mid yy \mid z$$

$$\text{e.g.} \quad 4 \ 8 \ 3 \ 9 \mid 25 \mid 7 \tag{4.6}$$

and we want to store this element, the last position z denotes the bus, yy denotes the bank and xxxx the element in a bank. For contiguous elements the last digit runs first, it denotes in the above decimal example one of the 10 busses 0 to 9. Then runs the bank number yy of each bus (100 banks 0 to 99), then the element number in the bank. In each cycle for the above example 10 elements would be transferred simultaneously by the 10 busses. So from the address directly the number of the bus by which the element must be transferred and the number of the bank in which it is stored or from which is loaded can be recognized.

4.5. Pipe Switch and Delay Register

Because the banks are connected only to certain busses a "pipe switch" must be inserted between the memory and the pipes so that an element of a certain bank can be directed to each pipe entry. A crossbar switch for such a large memory bandwidth

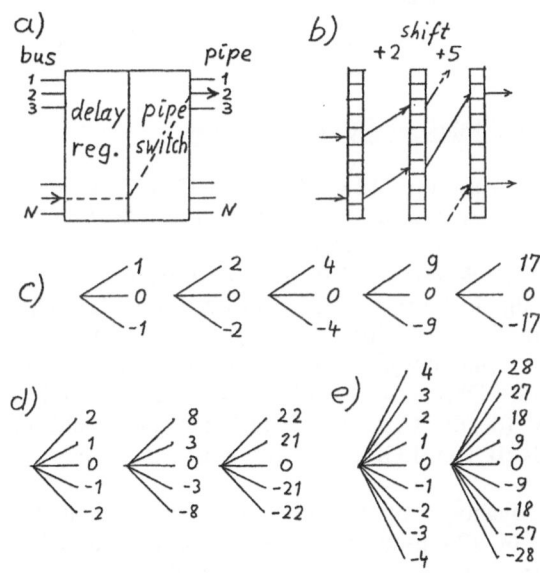

Figure 4.2. a) overview, b) principle of a 3-stage register transfer, c) 5 stages with 3 values, d) 3 stages with·5 values, e) 2 stages with 9 values (the value gives the corresponding shift).

would be too difficult, even if it could be built. Therefore we propose a pipelined register transfer, see Figure 4.2. In a store and forward manner in each cycle the contents of a register of 64 bits is transferred by a certain "shift" (cyclically) to another register, see Figure 4.2b. In Figure 4.2c is indicated how by 5 stages of 3 "values", in 4.2 d by 3 5-valued and in 4.2e by 2 9-valued stages the desired shift or switching can be obtained. A shift of 11 is, e.g., in 4.2 d obtained by $11 = -2 - 8 + 21$. The proposed stages allow all combinations of shifts - 32 to + 32 because of cyclic shift. Case 4.2d is to be preferred, but it is a question of integration if such complicated stages can be built. The pipe switch is one of the critical hardware design points of the CPVC.

From the above mentioned address scheme (4.6), that the last digit(s) denote(s) the bus, the shift is immediately given by

$$\text{shift} = \begin{cases} -i \text{ for } i \leq N \text{ ("up")} \\ N - i \text{ for } i \geq N/2 \text{ ("down")} \end{cases} \tag{4.7}$$

Here i denotes the bus number of the first element of a vector to be processed. This element must be "switched" to the first pipe, that has in the <u>addressing</u> scheme (4.6) the number 0 (zero). Note that the shifts are periodic. If the length n of the vectors to be processed is less than N, then in each cycle a different vector is processed and a different switching takes place. This switching is controlled by the control unit (CU) that guides its vector through the pipe, see below section 4.8.

In the worst case the three operands and the result of a vector triad (4.1) start in the same bank. Then the first operand must be read and delayed by 2 * bbt cycles and the second by bbt cycles. The result must be delayed by bbt cycles and then can be stored. Thus the load busses must have delay registers of the size 2 * N * bbt and the store bus of N * bbt. These delay registers are inserted between busses and pipe switch, see Figure 4.2a.

4.6. Building Blocks and Marketing Considerations

A supercomputer with 100 GFLOPS and 64 Gwords is a large supercomputer that not many institutions can afford. Therefore the CPVC should be composed of "building blocks": The smallest configuration would be 1 pipe (group) with 1.56 GFLOPS, 1 Gword (100 Mwords local, 900 Mwords extended). Then units of 1, 2, 4,..., 64 pipe groups should be marketed. But even 1.56 GFLOPS may be too large or expensive. Therefore the identically same CPVC should be built with a 12.8 nsec technology (CRAY-1 technology), yielding a series of computers from 156 MFLOPS, 100 Mwords to 10 GFLOPS, 6.4 Gwords. And for the minisupercomputer market a 128 nsec technology (nearly Convex C1 technology) should be used, yielding 15.6 MFLOPS, 10 Mwords to 1 GFLOPS, 640 Mwords.

The essential advantage of the use of these 3 technologies is, that the off--the-shelf technology of 128 nsec can be used for the "initial" design, the 12.8 nsec technology is readily available, and then the final technology of 1.28 nsec can be developed in the meantime. Above all the development and maintenance cost for hardware and software is distributed onto many units.

The extremely wide span of this identical overlapping family of (mini)supercomputers from 15.6 MFLOPS to 100 GFLOPS serves a correspondingly broad market, grasps the customers at their very beginning of supercomputing and allows then a continuous growth with increasing needs up to the limit of technology. If larger memories become affordable, an integration of 128 or 256 pipes in the same way as we discuss it for 64 pipes can be developed along with a further reduction of the cycle time. Present technologies will gradually develop to higher integration and shorter cycle time, but an essential breakthrough is expected if GaAs (gallium arsenide) memories with high integration will be developed in the next 8 to 10 years with a bbt of 0.5 nsec. The corresponding logic chips for the CPU will be developed in front of this memory technology.

Up to now much too much performance is lost by bad architectures. This is expressed by the last three performance-decreasing factors in (4.4). This loss in performance must be balanced by a shorter cycle time, i.e. by a more expensive technology. We think it much better, e.g., to build a 10 GFLOPS supercomputer with optimal architecture from a 12.8 nsec technology that delivers 10 GFLOPS sustained, than to build a supercomputer with 100 GFLOPS theoretical peak performance with a very bad architecture from a 1.28 nsec technology that delivers only 10 GFLOPS sustained. The only thing that counts for the user is the sustained performance.

4.7. Fail-safe System

There are hardware errors, there is maintenance. Therefore spare blocks should be used to replace "sick" blocks or to replace blocks for maintenance. The requirement for an easy replacement is that the address definition of a building block can be exchanged by the software, i.e. by the master control unit MCU, see below. The central hardware point for switching to other bus/memory blocks or other pipes is the pipe switch of section 4.5. Therefore the pipe switch should be designed for the possibility to include spare components. Also the software should be designed for error recovery. In an exception case an instruction retry for the faulty instruction (detected by parity control) should take place which means cancelling all instructions that have been started after the faulty instruc- tion. All possibilities for error recovery should be used. Thus a rather fail-safe system will result. This is essential for time-critical calculations (weather forecasting), or if the computer is integrated into CAD or CIM, or into SDI-type applications.

4.8. The Continuous Pipe

Now we explain why we call our computer a "continuous pipe" computer. In "usual" multipipe computers there holds the relation

$$n_{1/2,N} = N * n_{1/2} , \qquad (4.8)$$

where $n_{1/2}$ is Hockney's half performance length [11] for one pipe. This is the number of lost operations for each startup, depending on the type of operation. If the vector length n of the vector to be processed is just $n = n_{1/2}$, half of the processing time is wasted for startup and only half the time is "useful". If the operation is distributed onto N parallel pipes, $n_{1/2}$ is the N-fold of a single pipe. The value of $n_{1/2}$ determines the factor d in the performance formula (4.4). For a single pipe we have

$$\text{single pipe:} \quad d = \frac{n}{n + n_{1/2}} . \qquad (4.9)$$

If we use N parallel pipes, i.e. we use d in (4,5), we have

$$\text{N parallel pipes:} \quad d = \frac{n}{n + N * n_{1/2}} = \frac{n}{n + n_{1/2,N}}. \qquad (4.10)$$

It is indifferent if the task is executed in a bundle of N pipes or in N processors with one pipe, each. If $n = n_{1/2}$ for a single pipe or $n = n_{1/2,N}$ for N pipes, we have in (4.4) $d = 1/2$, i.e. the performance is halved. Thus for large N the value $n_{1/2,N}$ may have a decisive influence on the performance. If we assume $n_{1/2} = 20$ (excellent value), a bundle of 64 pipes has $n_{1/2,64} = 1280$. This means 1280 lost operations for each startup. This cannot be avoided for the first startup. But large computations always result from nested loops that result in sequences of vector operations. Question: Can we get $n_{1/2,N} \approx 0$ for sequences of vector operations?

In Figure 4.3 a the "classical" vector computer with a single control unit is depicted. The pipe or bundle of pipes is filled and cleared for each vector, resulting in $n_{1/2,N}$. But before a vector operation starts, i.e. data starts to flow, many preparing scalar operations must be executed. The addresses of the operands and of the result and the vector length must be delivered to certain registers, see the example of the vector triad in [13] for the IBM VF and CRAY X-MP. All these instructions contribute to the $n_{1/2}$. In Figure 4.3b the CPVC with <u>many</u> control units (CUs) is depicted, in the ideal case there are at least as many control units as there are stages of the load/execution/store pipeline. Each control unit "guides" <u>its</u> vector through the pipeline. The control units themselves are controlled by the master control unit (MCU). If we have sequences of independent and/or sufficiently long vectors, there is <u>no gap</u> between the different vector operations in the bundle of pipes. This is the basic idea of the CPVC.

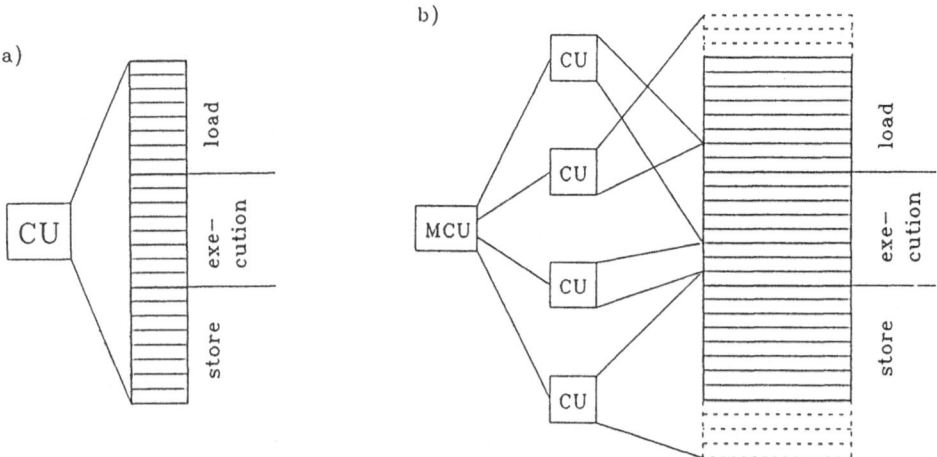

Figure 4.3. a) "classical" vector computer with a single control unit, b) CPVC with many control units, CU = control unit, MCU = master control unit.

An essential effect of the many CUs is that these units have "time" to prepare the operations. They form a queue and can immediately start their vector operations if the preceding vector operation frees the first stage of the pipe or if the condition to start the operation is met. The $n_{1/2}$ starts in this case only if the first data starts to flow. All preparations have been overlapped with preceding operations because the corresponding CU has to deal only with its own vector that it guides through the pipes. Thus the $n_{1/2}$ or the $n_{1/2,N}$ for the first startup is also considerably reduced so that $n_{1/2}$ = 20 may be a reasonable value for the first startup (e.g. 3 cycles for memory to pipe, 4 + 5 + 4 cycles for the addition/multiplication/addition pipeline, see below, 3 cycles for pipe to memory, resulting in 19 cycles). These lost cycles result only for the first startup of a sequence of vector operations. In a sequence the operations are executed without lost cycles because they follow each other without a gap, i.e. without lost operations.

But for 64 pipes vectors are shifted in "segments" of 64 elements through the pipe, these are the "stages" in Figure 4.3b. In the last stage of a vector operation may be between 0 and 63, in the "mean" $n_{1/2,N}$ = (N - 1)/2 lost operations. Thus $n_{1/2,N}$ = 0 cannot be obtained even in a sequence, but the above value is the best we can get, this means $n_{1/2}$ ≈ 1/2 for a single pipe which expresses the fact that there is 50 % probability for a dummy operation of a pipe in the last segment of a vector operation.

4.9. Vector Dependencies

For $n_{1/2}$ = 20 stages of the pipe and N = 64 the "volume" of the bundle of pipes is 20 * 64 = 1280 elements (= $n_{1/2,64}$). Therefore there is a "critical" vector length n_c = 1280. For a vector with $n > n_c$ elements no precautions are necessary. But for $n \leq n_c$ we have vector dependencies for dependent vector operations. The expression

$$a_i = b_i + c_i * d_i - (e_i * f_i + g_i) * h_i \qquad (4.11)$$

can be executed as 3 vector triads. Then, e.g., $(e_i * f_i + g_i)$ is an intermediate vector that must be out of the pipe before it can be reused as operand in a following operation. Or in the matrix-vector multiplication in diagonal form we have an operation of the type

$$c_i^{(s)} : = c_i^{(s)} + d_{i,\nu} * r_i^{(s)} \qquad \text{for all } \nu , \qquad (4.12)$$

where s means shifted part. The vector $c_i^{(s)}$ of the left hand side must be out of the pipe before it can be used as operand on the right hand side for the next ν . Thus the problem are the short dependent vectors ($n \leq n_c$). In this case we should use a register file to avoid the load/store part of the pipeline, see the "data flow graph" of Figure 4.4. The register file is only for intermediate results, its volume should be at least 4 times the volume of the pipes. For operations with the register file the critical vector length is reduced to the volume of the combination pipelines (see below) and dependent vectors up to this length can be processed via the register file without loss. For dependent vectors smaller than the volume of the pipes (N * stages of combination pipeline) lost cycles arise for waiting until a dependent operand is out of the pipe. The existence of the register file means that the compiler must generate two codes in those cases, where at compiletime the decision cannot be made if the operation is memory-to-memory or if it uses the register file. Then at execution time the MCU must select either of the codes and when this cannot yet be decided at the moment of the instruction issue, the memory-to-memory code must be selected.

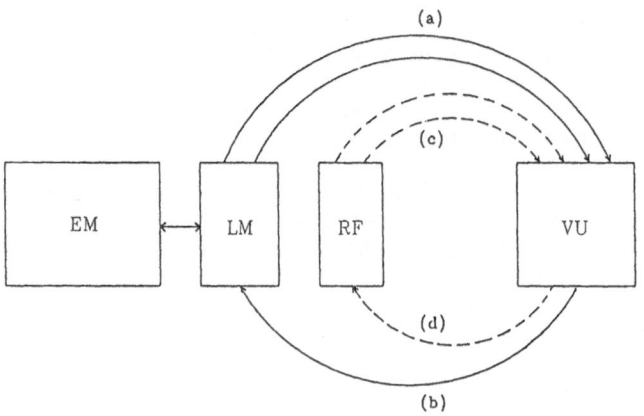

Fig. 4.4. "Data flow graph" of the CPVC. (a), (b) is for initial/long/final vectors, (c), (d) is for short intermediate vectors, EM = extended memory, LM = local memory, RF = register file, VU = vector unit.

For a "contraction" operation like (4.12) a "vector wheel method" could be devised, where several "copies" of c_i are turning around (for short vectors via the register file) and are finally contracted to a single vector. It should be mentioned that the efficiency of the CPVC drops drastically if n<N. Then idling of pipes is unavoidable. But this holds for all parallel designs. If a problem with dominantly short vectors must be treated it is either a small problem and can be treated efficiently only on a vector computer with a single or only a few pipes (small N), or if it is a really large problem then it is not vectorized appropriately, i.e. the innermost loop is not the appropriate loop.

The CPVC "normally" is a memory-to-memory computer. But for short dependent vectors that use the register file for intermediate results, it is similar to a register-to-register computer, but it avoids the unnecessary load/store operations for registers because the register file is used only for intermediate results, see Figure 4.4, i.e. without "explicit" load/store operations. This is only possible due to its 4-word per cycle and pipe memory bandwidth.

4.10. Combination Pipeline

The "execution" part in Figure 4.3b is for triadic operations the combination of addition and multiplication. But there are operations a * b + c where the result of the multiplication pipe is delivered to the addition pipe, and operations (a + b) * c where the result of the addition pipe is delivered to the multiplication pipe. If we have in a pipe group or combination pipe several vector operations, guided by their control units, see Figure 4.3b, we could not mix up freely these two types of operations for a synchronous execution. Therefore we propose a combination pipe of addition/multiplication/addition, see Figure 4.5.

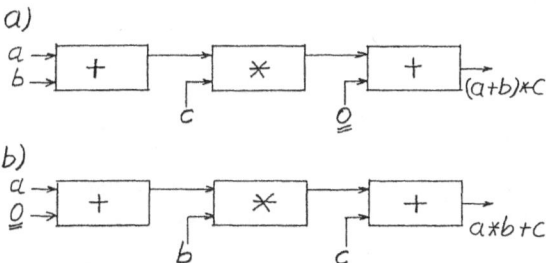

Figure 4.5. Combination pipe.

The disadvantage of this combination pipe is that it is "longer", i.e. needs more cycles, than a combination pipe composed only from multiplication and addition. But this counts only for the first startup of a sequence of vector operations, it does not count in the sequence. The essential advantage is that we have a fixed pipe length for the execution part which is necessary for a synchronous mixing of combined operations of different kinds. Because of the synchronization even the simple operations a + b and a * b will be executed in this combination pipe by feeding neutral elements 0 of addition and 1 of multiplication into the corresponding functional units. Such a combination pipe could also execute the expression (a + b) * c + d with triple vector speed if one of the operands is a scalar, but one should not count because of this "possibility" the peak performance by this very rare operation. Note that a + b + c can also be executed with supervector speed. Special procedures can be developed for contracting operations, e.g., scalar product.

4.11. Scalar Speed

If we assume a bundle of 64 pipes for vector operations, then there is a misbalance to the scalar speed, which means that we need a very high degree of vectorization in order to get a good speedup from scalar to vector execution, see [1,11]. This is known as Amdahl's law. There is always a remainder of scalar code. But also the last stages of an algorithm with shrinking vector length should be better executed in scalar mode if we are below the breakeven length where scalar execution is faster than vector execution. In the LU-decomposition of a matrix the vector length shrinks to 1 and thus the last stages are below the breakeven length. Then we can unroll (completely) the innermost loop and get scalar operations that can be executed in parallel. If we have a scalar expression

$$a = b * c + d * e + f * g + h * i \qquad (4.13)$$

there are 4 "independent" multiplications and 2 "independent" additions. With parallel addition and multiplication we could execute (4.13) in 3 "chimes" instead of 7 "chimes" for sequential execution. Therefore we propose several scalar arithmetic units that are controlled by scalar control units. Then we have the configuration of Figure 4.6: The master control unit controls several scalar and vector control units. The multiple scalar control units allow above all a parallel "preparation" of independent and

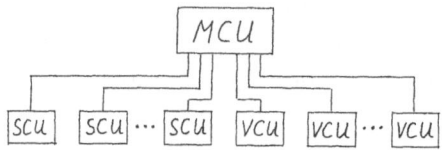

Figure 4.6. Overview of the CPVC, SCU = scalar control unit, VCU = vector control unit, MCU = master control unit.

in some sense also of dependent scalar operations which increases significantly the "total" scalar speed. In a computer with very fast vector unit we must use every possible chance to make the scalar part as fast as possible. This can be made only by several parallel scalar units.

4.12. Program Execution

If we have several scalar and vector control units that might be active at the same time, this means that we have parallel execution of independent operations. But we must be careful to synchronize <u>dependent</u> operations. This synchronization must be detected by the compiler from a data dependency graph [12], but it must be executed by a corresponding hardware design. We propose a number n_s of synchronization flags that have values 0 and 1. Each operation may depend on such a synchronization flag and can start only if its value is zero. Each operation can "set" several synchronization flags of other operations that depend on this operation, it sets 1 as long as it executes, and sets 0 when it terminates.

Because each operation is executed or "guided" by a CU, either a VCU or a SCU, see Figure 4.6, "operation" is synonymous with CU. This means that the synchronization flags control and/or are controlled by the CUs. Because the CUs are "intelligent" units, they can recognize when the first element of an operation is back in the memory. At this moment the synchronization flag can already be set to zero, a dependent operation can start and use this result.

In the same sense we propose a number n_c of condition flags that have three values: a condition flag has value 0 as long as the condition is still open, it has value -1 if the condition is false, then the operation under the control of this condition is "superfluous" and terminates, and it has value +1 if the condition is true, then the operation starts. The condition flags are useful for the "preparation" of alternatives in different CUs, that the operations can start as soon as the condition is set. All these flags serve for the creation of a high parallelism also for scalar operations.

The "program" is executed in the MCU, see Figure 4.6, that "creates" instructions and passes them to the SCUs and VCUs. Outer loops are expanded in sequences of vector operations that are passed to and executed by the VCUs. The instruction set must care for such "mighty" operations of the SCUs and VCUs that correspond to several operations on "classical" vector computers. The instruction format should be similar to VLIW (very large instruction word), e.g. the most complex vector operation

$$e_i = (a_i + b_i) * c_i + d_i, f_s, f_c, sf_{s1},....sf_{sm}, \qquad (4.14)$$

(one operand must be a scalar) must be <u>one</u> instruction from the MCU to the VCU, where f_s, f_c are synchronization and condition flags that control this operation, and sf_{s1},..., sf_{sm} are for setting m synchronization flags for synchronizing other operations. An operation like (4.14) must be passed by operation code, vector length, starting addresses of the operands in main memory or register file. Similar mighty instructions must be issued to the SCUs.

The MCU must be able to issue every cycle such mighty instructions like (4.14). This is only possible if the MCU has also mighty instruction pipelines that allow the processing of instructions and even loops so that every cycle an instruction can be delivered to a VCU or SCU, see Figure 4.6. Thus the MCU has internally several units that operate themselves in parallel. The design of the MCU is also one of the critical points of the CPVC. But this demonstrates that the CPVC is no longer a v. Neumann computer but a really <u>new architecture</u> that is designed to treat parallelism on the <u>hardware</u> level.

4.13. Data Transfer Operations

The necessity and importance of data transfer operations has been mentioned in section 2.2. Masked data transfer operations like pack, unpack, merge profit from the continuous flow of data in the CPVC like the arithmetic operations. They should be executed in the combination pipe, using the three load pipes and the store pipe for operands, mask and result. Also dyadic masked operations could be executed in the combination pipe, using two loads for the operands and the third load for the mask. Then the store is under the control of the mask. The generation of masks should take place also in the combination pipe. Thus even between different types of operations no gaps occur in the combination pipe.

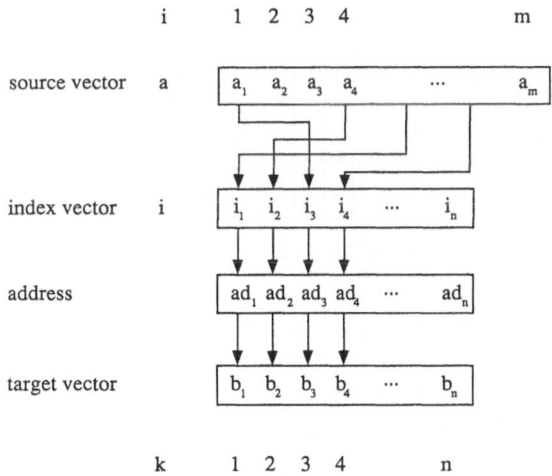

Figure 4.7. Illustration for gather.

A new problem are the indirect addressing operations controlled by an index vector, in Fortran 77 notation:

gather: b (k) = a(i(k))

scatter: a (i(k)) = b(k)

(4.15)

Such operations are essential for all types of unstructured data, e.g. the FEM method.

Because of the importance of these operations we explain in more detail how for the special memory structure of Figure 4.1 data can be selected by an optimized priority-controlled two-stage gather nearly with vector speed. The situation is depicted in Figure 4.7. The elements a_i to be selected are stored in memory banks that are accessible only by a distinct bus, see Figure 4.1. If a bank has been accessed, it can be accessed again only after bbt cycles. Therefore the access must be optimized for the busses and for the banks. The basic idea how to do this is illustrated in Figure 4.8 for the example of the selection of 12 elements for a CPVC with N = 4 pipes. The distribution of the elements onto the banks is depicted by their indices k. A bank with p elements gets a priority p. In a first step N elements with highest priority are selected (gathered) and stored in an intermediate register. In the next step these elements are stored (scattered) in their correct position into a special register file. In the same step the next N elements are selected and are stored in the following step into the intermediate register. But banks that have been accessed must be excluded for bbt cycles. This is continued until all elements are gathered.

Now we explain in more detail how this procedure can be cast into hardware. The reader not interested in these details may skip to section 4.14. We illustrate the procedure for the decimal addressing scheme (4.6), with 10 busses and 100 banks for each bus, see Figure 4.9. The priorities are stored in a priority stack whose columns are the bank stacks (BSs), one for each bank. An availability bit is "1" if the bank can be accessed, it is "0" if the bank is busy. If an element is selected from a bank, during bbt cycles the availability bit of this BS is 0 and then again is set to 1. The priority counts the number of elements to be selected from that bank. In the BS are stored the k's of these elements. Note that for the address scheme (4.6) the last digit

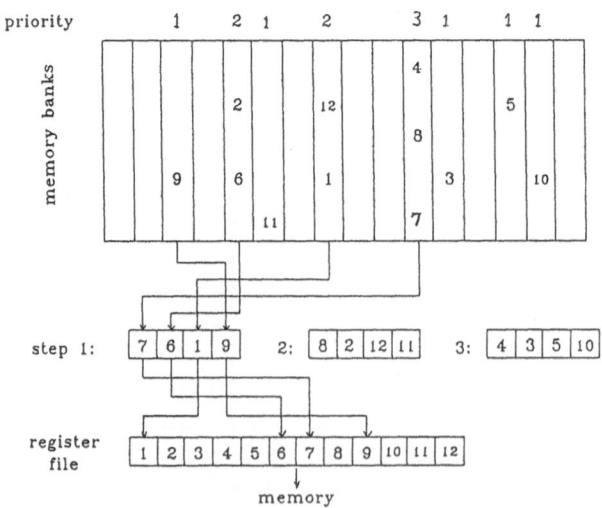

Figure 4.8. Illustration for a two-stage gather for N = 4.

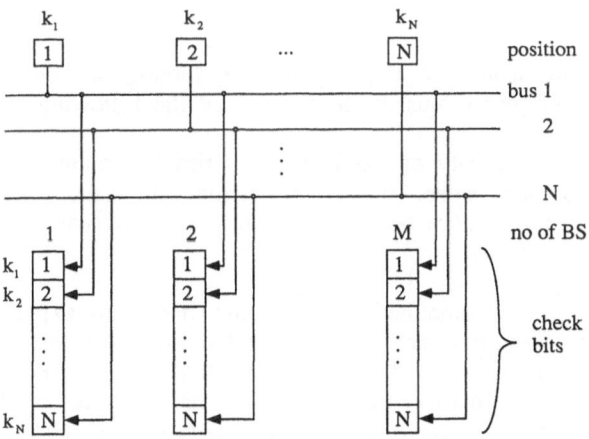

	(N-1)			
bus	0	1		9
bank	0 1 2 3 4 5 .. M_b-1	0 1 .. M_b-1	...	0 1 .. M_b-1
avail. bit	0 0 1 1 1 0 0	.		.
priority	3 0 1 2 4 0 1	.		.
	k k k k k	.		.
k's	k k k			
	k k			
	k			

Figure 4.9. Priority stack with the bank stacks.

denotes the bus and the next two digits the banks of that bus. The execution of the gather/scatter operations is made by a special gather/scatter unit that has its own register file. For the processing of a gather the following 4 steps are executed:

Step 1: A special instruction reads sections of length N of the indexvector i, composes the addresses ad_k of these elements, see Figure 4.7, and stores them in an address register file. If we assume a maximal length m_r (e.g. 1024) of the register file and address register file, this is continued until m_r or the remainder of n. Thus a vector of length n is stripmined with length m_r.

Step 2: As soon as N addresses are stored in the address register file, the filling of the BSs starts, again parallel for N elements, see Figure 4.10. For N positions the next k's are selected and for these k's the addresses ad_k are selected from the address register file. By these addresses with a hardware circuit of Figure 4.10 check bits in N positions of the M banks are set. Then for those check bits that have been set the corresponding k_j's are stored in the BSs of Figure 4.9 and the priority is increased correspondingly. This

Figure 4.10. Illustration for the filling of the BSs.

process must be executed in a pip elined mode so that in each cycle N k's can be processed. But each BS has only a limited capacity, e.g. 8 or 16 or 32 elements. So it may happen that one BS is "filled up". Then the process of filling must be stopped, the current section up to the processed indices must be terminated and the not yet processed indices are processed in the next section of length m_r.

Step 3: As soon as N elements are stored in the priority stack, parallel to the filling, but asynchronously, the emptying of the priority stack starts. For each of the N busses the k with highest priority, whose availability bit is on, is selected. This guarantees that the banks with most elements are accessed immediately when their bbt is over. If for a bus no k is available, the bus idles in that cycle. For the selected k's the addresses ad_k are selected. For these addresses the elements $a(i(k))$ are fetched from the memory, using one of the load bus groups, and are stored in an intermediate register. Then the elements of the intermediate register are stored according to their k's in the register file as illustrated schematically for N = 4 in Figure 4.8. Again this procedure is executed in a pipe-lined mode.

Step 4: If the whole priority stack is empty, i.e. all BSs are processed, the register file of length m_r, that now contains a section of the elements b_k (or the remainder), is dumped to the main memory by the store bus. These are now contiguous elements.

By the pipelining principle, in the optimal situation that no bank conflicts occur, for the first section of m_r elements two cycles per element are needed: one for the selection and one for the storing, neglecting the startup time. All following sections need only one cycle per element because filling and storing of the register file can be executed in parallel if two alternating register files are used. Then we have nearly vector speed for the gather operation. Bank conflicts force some busses to idle and this reduces the performance correspondingly. But this is unavoidable. If we have m elements in the same bank we need at least m * bbt cycles for the selection of these m elements. The above procedure minimizes the cycles for the selection of the m_r elements of a section of the vector b.

In a completely analogous manner an optimized two-stage scatter can be executed. The steps 1 and 2 for the building up of the priority stack are exactly the same as for the gather. But the steps 3 and 4 must be inverted:

Step 3': The b_k's are loaded to the register file (contiguous elements). The whole section of length m_r must be loaded before the following step can start.

Step 4': Following the priority of the BSs the optimal elements are selected from the register file and stored in an intermediate register. From there they are loaded by "their" store bus to the memory. This runs free of bank conflicts.

The procedure for a pipelined gather/scatter has been explained with so much detail because these operations are decisive for the flexibility and overall performance of the CPVC. At first sight it does not seem possible to select with nearly vector speed elements from a memory with the access structure of Figure 4.1. The corresponding hardware unit for gather/scatter operations is also one of the critical components of the CPVC.

4.14. Software

For the user the CPVC is a large SIMD monoprocessor whose internal parallelism is transparent to the user. On the user level all software problems are solved. He has to care only for optimal vectorization, i.e. he has to choose appropriate data structures for the pipeline principle. But he has not to care for special bottlenecks like the cache size of the IBM VF or the memory bottle-neck of all existing supercomputers. Because he must not adapt his program to the particularities of a special computer, his investments in application software are preserved in the future.

The compiler also has not to care for special bottlenecks. But it should detect by the data dependency graph [12] independent operations that can be executed "in parallel", i.e. can be together in the bundle of pipes or can be executed independently on the scalar units. The compiler should also detect the dependencies so that the MCU can pass the corresponding synchronization information to the CUs.

Such a supercomputer whose main purpose is number crunching should have a rather simple operating system only with batch functions. There should be a very high priority job class for quick answer, but no direct interactive function. Disks should be used only for short-time storage. All long-time file management and the service for slow periperals should be done by a front-end computer, where also the compiler and tuning tools are installed. Also all networking should be done by the front-end. Thus the supercomputer is freed from all non-numbercrunching problems. This has not only the effect that the expensive pipes and memory can be used optimally for their proper purpose with minimal idle time, but also that the instruction set can be designed specially for the pipelining instructions. Therefore we propose that a front-end computer should be automaticaly included in the CPVC for all non-numbercrunching problems.

Today the interaction between a supercomputer and many graphical workstations is an essential feature of supercomputing. The operating system should be designed so that an easy communication between a job of the supercomputer, running in a special batch job class, and a workstation can be established. The basic principle in such a communication must be that the workstation has to wait and not the supercomputer. Thus the throughput of the supercomputer is least disturbed by the interactions.

5. CONCLUDING REMARKS

If we need parallelism of vector pipes we should organize this parallelism in a user-friendly way, like in the CPVC. In the CPVC all problems of parallelism are shifted to the hardware, where they belong to and can be solved there, and not to the user. The cycles that are lost on the hardware level can never be regained on the software level. The CPVC has features of a vector computer, of a hierarchical computer (MCU, CU, pipe), of a data-flow computer (the pipes are kept continuously busy by the special design) and of a VLIW (very large instruction word) computer (mighty instructions that are executed in parallel).

The CPVC is not a pure SIMD computer. A single instruction, i.e. a vector operation, acts on multiple data, i.e. on the elements of a vector. But in the bundle of pipelines many such different vector instructions can be active at the same time and act on different data. In this sense it is also a MIMD computer. The essential feature is that it is a synchronous computer. All memory accesses for that memory bandwidth are synchronous. This is the only way to keep the data continuously flowing, i.e. to keep the pipes continuously busy. But this synchronization of all operations results in extreme demands for the hardware design and this is the price to be paid. There is no other way to manage the large amount of data that is moving around in the CPVC, 100 GFLOPS need that the corresponding operands and results are flowing through the computer. The "classical" vector computer of Figure 4.3a is still too close to the v. Neumann type of computer. The step to the CPVC of Figure 4.3b is the only way towards a user-friendly parallelism, a parallelism that solves the problems optimally on the hardware level.

The CPVC comes with less effort of the user much closer to the theoretical peak performance than any other architecture with parallel vector pipes or parallel vector processors. Its sustained rate is for well designed programs for the solution of large problems close to the theoretical peak rate because the number of lost cycles is minimized. We want to stress again that the main purpose of this paper is to demonstrate that and how we could step to an architecture that makes the parallelism transparent to the user, i.e. that is user-friendly.

The only competitor for the CPVC in the domain of large supercomputers is the message passing system, a "true" MIMD computer. It would obtain the necessary memory bandwidth for each processor in its local memories. But just for this reason a central resource management can <u>never</u> be obtained, there is no program switch between different programs in a multiprogramming mode with time slicing for the CPU power. Thus a message passing system remains always a PC. It is composed of v. Neumann computers (with insufficient memory bandwidth) and thus remains also a v. Neumann computer in its core. The problem of coordination of the CPU power is shifted to the software level and thus is shifted to the user. The software depends strongly on the communication network of the message passing system and thus is architecture-dependent. This is <u>not</u> a user-friendly architecture. Such a computer has its merits and its (eventually) better price-performance relation only as a special purpose computer and there it should be used with advantage. It would be wasted money to use a sophisticated architecture like the CPVC for a fixed class of problems, e.g. radar image processing. If we need a certain performance in MFLOPS or GFLOPS we have to ask: What is the best and, including the software cost, the most economic architecture that can deliver this performance. And don't forget: It is only the <u>sustained</u> performance that counts, not the theoretical peak performance.

6. WEAK POINTS OF PRESENT SUPERCOMPUTER ARCHITECTURES

In the light of the ideas of the CPVC we briefly revisit the weak points of present supercomputer architectures.

<u>Memory bandwidth:</u> Many processors have only 1 word per cycle and pipe group, the best are the CRAY X-MP, Y-MP with 3 words. Four words are needed for a flexible use of the addition and multiplication pipes.

<u>Vector registers:</u> They are used to bridge the narrow memory bandwidth. This leads to the "sectioning" of vector operations with new startup after each section.

<u>Memory size:</u> The main memories are (usually) too small and restrict the problem size. The available extended memories have a too narrow bandwidth to be real extensions of the main memory. One word per cycle and pipe would be needed for the extended memory.

Large multiprocessors:	If they are used as MIMD computers by multi-, makro-, micro-, auto- tasking the problem must be broken (artificially) into pieces.
Parallel computers:	The problem and the data are distributed on the software level to the processors and broken (artificially) into pieces. The software is highly architecture-dependent because it depends on the communication pattern. The problems are shifted to the user.

An example of a supercomputer architecture is the CRAY-2 of the University of Stuttgart, measured in November 1986, see [1]. The vector triad for n = 10 000 delivers for one processor as a mean value of 10 measurements (wide scatter) a performance of 51.4 MFLOPS. This is the real world. The theoretical peak performance is 487 MFLOPS. This can never be obtained. The measurement of the matrix multiplication 512 x 512 by an assembler library routine delivers 311 MFLOPS. This is the exception of the repeated contracting linked triad. If we accept the general triad as representative operation, we could obtain the 51.4 MFLOPS with a 38.9 nsec technology (nearly CONVEX C 200 technology) instead of an expensive 4.1 nsec technology, if we had an architecture without lost cycles. This illustrates what we ment when we wrote above that it is better to build a 10 GFLOPS computer that delivers 10 GFLOPS sustained than to build a 100 GFLOPS computer that delivers 10 GFLOPS sustained, because the latter must pay for the bad architecture by a much more expensive technology. Unfortunately we have to live presently with such shortcomings and to pay good money for bad architectures.

7. REFERENCES

1. W. Schönauer, *Scientific Computing on Vector Computers*, North-Holland, Amsterdam (1987).
2. W. Schönauer, "Why I like Vector Computers", Interner Bericht Nr. 35/89 des Rechenzentrums der Universität Karlsruhe, Karlsruhe (1989).
3. W. Schönauer, "Why I like Vector Computers", in *Supercomputer '89*, H.W. Meuer, ed., Informatik-Fachberichte No. 211, Springer (1989), 119-146 (This is a reprint of [2], excluding the Epilog).
4. P. Sternecker and W. Schönauer, "A Finite Element Kernel Program with Optimal Data Structures for Vector Computers", Interner Bericht Nr. 34/88 des Rechenzentrums der Universität Karlsruhe, Karlsruhe (1988).
5. P. Sternecker and W. Schönauer, "VECFEM: Effective computation of large FEM-problems on vector computers", in *Application of Supercomputers in Engineering: Algorithms, Computer Systems and User Experience*, C.A. Brebbia and A. Peters, eds., Elsevier, Amsterdam (1989), 41-51.
6. J.J. Dongarra, ed., *Experimental Parallel Computing Architectures*, North-Holland, Amsterdam (1987).
7. *Parallel Computing*, vol 7, No. 3, (Sept. 1988), 263-499.
8. W. Giloi, "SUPRENUM: A trendsetter in modern supercomputer development", in [7], 283-296.
9. A.H. Karp, "Programming for parallelism", Computer, 20:43-57 (1987).
10. D.J. Kuck, E.S. Davidson, D.H. Lawrie and A.H. Samek, "Parallel supercomputing today and the CEDAR approach", in [6], 1-23.
11. R.W. Hockney and C.R. Jesshope, *Parallel Computers 2*, Adam Hilger, Bristol (1988).

12. D.J. Kuck, "Automatic program restructuring for high-speed computation", in *Conpar 81*, W. Händler, ed., Springer, Berlin (1981), 66-84.
13. W. Schönauer and H. Häfner, "Vectorization, optimization and supercomputer architecture", in this book.

THE USE OF TRANSPUTERS IN QUANTUM CHEMISTRY

U. Wedig, A. Burkhardt, and H.G. von Schnering

Max-Planck-Institut für Festkörperforschung, D-7000 Stuttgart 80, F.R.G.

QUANTUM CHEMISTRY AND COMPUTER

Quantum chemistry is one of the areas in natural sciences that is most closely related to the development of computers. The several methods to approximately solve the basic equations describing the electronic structure of atoms, molecules and solids (e.g. the Schrödinger equation), all have in common the large amount of computational expense. For ab-initio methods, that is for methods that do not use parameters fitted to experimental data, this expense is proportional to about n^3 (density functionals), n^4 (Hartree-Fock) or even n^5 (configuration interaction (CI)), where n is a quantity related to the accuracy of the calculation and to the size of the problem treated. Looking at this n dependency we understand, why very powerful computers are necessary to envisage reasonable problems.

The present generation of modern supercomputers consists of a small number of sequential processors, supplemented by special hardware to execute vector operations very efficiently. The processors are connected to a global memory, allowing the communication between tasks running in parallel. Algorithms with large contiguous data structures and preferably no branches in the inner loops should be used for vector computers.

Programs for the transformation of the two electron integrals from an AO to a MO basis or direct CI programs nearly exploit the peak performance [1]. However, this is not true for other algorithms used in quantum chemistry. One of the most important one is the Hartree-Fock method (cf. [1,2,3]), which will be discussed later.

PARALLEL COMPUTER ARCHITECTURES OR WHY WE USE TRANSPUTERS

To overcome the problem of poor vectorizability, and in addition to achieve new dimensions of performance, parallel computation has to be investigated. Experience with the parallelization of quantum chemical software was published, besides our work [4], by Colvin [5] (cf. [6]) and by Clementi's group [7]. Different Hartree-Fock programs had been implemented on loosely coupled arrays of processors (IBMOL6 [8], HONDO [9] and GAMESS [10]).

Scientific Computing on Supercomputers II
Edited by J. T. Devreese and P. E. Van Camp
Plenum Press, New York, 1990

Starting with parallel computing one has to keep in mind the large variety of existing parallel architectures. They differ in the number and the capabilities of the processors, as well as in the way of interprocess communication. Machines with global memory like the above mentioned vector supercomputers are restricted to few processor nodes, due to limited memory bandwidth and to the growing complexity of controlling the access to the same memory locations. Massively parallel systems can be build from processors, each disposing of his own local memory. The nodes may be connected by a common bus or by several multi-dimensional networks (see Figure 1). The hardware concepts are completed by various programming techniques. Single instructions (SIMD) or complex programs (MIMD) of different granularity may be distributed among the nodes. This large variety is elucidated by a quotation, summarizing the results of a workshop on parallel algorithms and architectures [11].

We can call it 'parallel processing', but we don't know how to do it. We do need experiments, however, to suggest possible new directions.

Considering this statement, we surely cannot proclaim transputers being the only solution. However, they are an efficient and good value starting-point for the experiments needed. The transputer [12] is a computer on one chip, especially designed to be an element in a massively parallel system. The components are shown in Figure 2. Besides the CPU, the chip contains a floating point unit, fast local RAM, and an interface to external memory. Four other transputers can be connected to the independent serial link interfaces to build up larger transputer networks. Parallel to the transputer, INMOS developed the programming language OCCAM [13]. OCCAM enables to program concurrent processes, communicating through unbuffered, unidirectional channels. With OCCAM, the hardware facilities of the transputer can be utilized very efficiently. For more detailed information on both, the hardware and OCCAM, we refer to some articles of Hey [14,15].

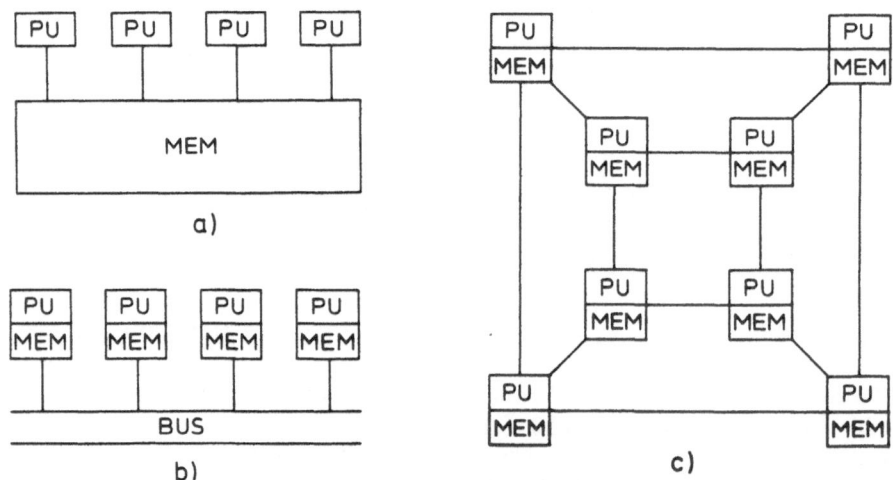

Figure 1.   Connection of processing units (PU) in parallel systems.
a) global memory
b) local memory, bus system
c) local memory, multidimensional network

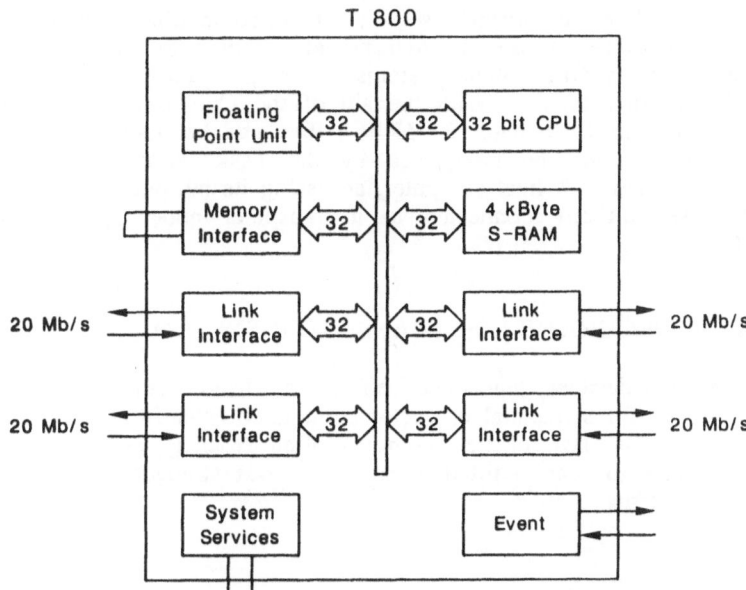

Figure 2. The INMOS transputer T800.

PROGRAMMING ENVIRONMENT FOR TRANSPUTER SYSTEMS

Two different types of user interfaces for transputer systems are available. The first one is the transputer development system TDS, supplied by INMOS. We used a modified version of TDS called MultiTool [16]. The second type are UNIX-like, distributed operating systems like Helios [17].

TDS, MultiTool

TDS consists of a folding editor, the compiler, the configurer, and several utilities. The processes run on a single node, which is connected to a host computer (PC, SUN workstation ..) via a link adapter. A server process is located at the host. It communicates with TDS, handling all the requests to the hosts peripherals. In general, TDS doesn't use the other nodes of the transputer network, so they are free for application processes. TDS is the ideal environment for writing and executing OCCAM programs.

Helios

Helios is a distributed operating system for transputer systems. On every node runs a nucleus, providing the basic functions necessary to control the processor. It consists of the Kernel, the system Library, the loader and the processor manager. The message passing algorithm for the basic interprocessor communication is implemented in the Kernel. Other system services are based on the Client-Sever model. Server tasks, which are located somewhere in the network, wait for messages from application tasks that request for some operation or reply.

The UNIX-like command set, a Unix-compatible library and standard programming languages form an environment familiar to many users. Therefore, transferring programs from other systems is easier. Parallelism is not defined in programs (tasks) themselves. With the aid of the Component Distribution Language (CDL), the job control language for Helios, they can be defined as components of a Task Force, which will be interpreted by the Task Force Manager. The Helios network is connected to a host computer, accessing its peripherals via a server process. To handle disks directly connected to a transputer node, special file servers are available (e.g. [18]).

DEVELOPING PROGRAMS FOR TRANSPUTER SYSTEMS

In the last decades, enormous effort has been spend to optimize quantum chemical software for sequential or vector computers. Nearly all of the code is written in FORTRAN. The size of the programs often reaches several 10 000 statements. This fact we have to bear in mind, if we think over strategies to implement programs on transputer systems.

Programming in OCCAM

This is surely the most suitable way to create and control concurrent processes on a network of transputers. However, rewriting the most important quantum chemical programs in OCCAM would require many man years of work, compensating the favorable price/performance ratio of massively parallel systems. This disadvantage is of consequence all the more, because the future of OCCAM cannot be foreseen. Communication and configuration elements in OCCAM compete with the message passing and task controlling mechanisms in distributing operating systems like Helios. The most important vendors favor this latter way. Therefore, they do not support OCCAM in a manner necessary to make it a standard in programming parallel algorithms.

Farming

Using the farming concept [15,19], we can easily implement existing programs on a transputer system. Identical sequential tasks may be distributed in the network. A master (farmer) process provides every worker task with input as its processor runs idle and collects the output for further processing.

Farming on the program level

If the worker tasks are complete programs, we can use existing FORTRAN code practically without modifications. Routing of the data and load balancing can be controlled by small OCCAM programs or more easily with the aid of CDL under Helios. The transputer cannot manage virtual memory. All processes running on one node have to fit into its physical memory. That is why farming on the program level requires more expensive nodes.

For economical reasons we cannot equip every node with disks. So we have to manage the access to remote devices. Under Helios this is done by the message passing routines in the Kernel. In a later section we discuss our experience with Helios in this connection. As each program runs on one node, its execution time will not be reduced, compared to a single processor machine, although the throughput of the whole system will be much higher.

If we want to decrease drastically the execution time of a whole program, programming work is unavoidable. We have to decompose the code into independent tasks that can be distributed into the network. If we can use modules or subroutines out of existing programs, the work is restricted to the insertion of the message passing and load balancing algorithms. Provided a farm structure is suitable, again this is rather easy. The same procedures as mentioned above can be used.

The worker tasks in general require less hardware resources. Farms on the subroutine level can run on cheaper networks reaching nearly the throughput of farms on the program level. In a farm we do not predetermine which and how much data is processed on a node. Therefore we can adapt the distributed program by a simple reconfiguration to different networks which even may contain processors with varying performance.

A DIRECT SCF PROGRAM

To calculate the electronic wavefunction in the Hartree-Fock approximation we have to solve a generalized eigenvalue problem

FC = eSC

The single particle functions (molecular orbitals) used to construct the total wavefunction are linear combinations of n basis functions. The matrix **C** contains their expansion coefficients. The elements of the Fock matrix **F** for a molecule with closed electron shells are calculated as follows.

$$F_{pq} = h_{pq} + \sum_r^n \sum_s^n P_{rs} \, (2g_{qrps} - g_{qrsp})$$

As the elements of the density matrix P_{rs} are functions of the expansion coefficients, the Hartree-Fock equations have to be solved iteratively (**SCF: Self Consistent Field**).

In most cases, cartesian or spherical gaussian functions are used as basis functions and the one- (h) and two electron integrals (g) are calculated by analytical formulas.

In Figure 3 we show the structure of the direct version of a SCF program [20,21,22,23]. A discussion of the differences to the conventional procedure and the benefits of the direct variant can be found in the cited papers.

The most time consuming step in every iteration is the calculation of up to $\approx n^4/8$ two electron integrals. To save arithmetic operations, basis functions with the same angular quantum number, the same exponent and the same location are collected in shells. Batches of integrals, comprising of all two electron integrals of a quadruple of shells, are calculated in one process. The size of batches can differ from one other by some orders of magnitude (4 different s-shells: 1 integral; 4 different d-shells: 1296 integrals (cartesian gaussian functions)).

Batches can be processed independently from the others, so that the calculation of the two electron integrals can be executed in parallel. Load balancing problems because of the varying granularity of the single processes are prevented by the farming mechanism.

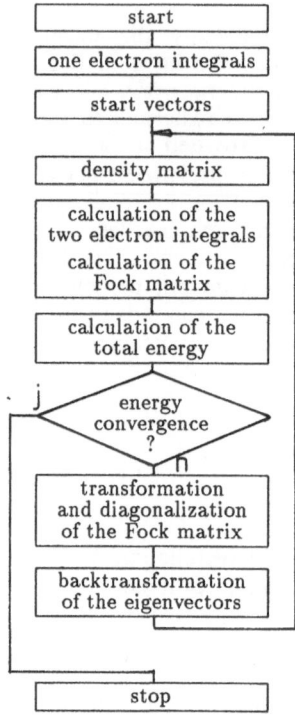

Figure 3. A direct SCF program.

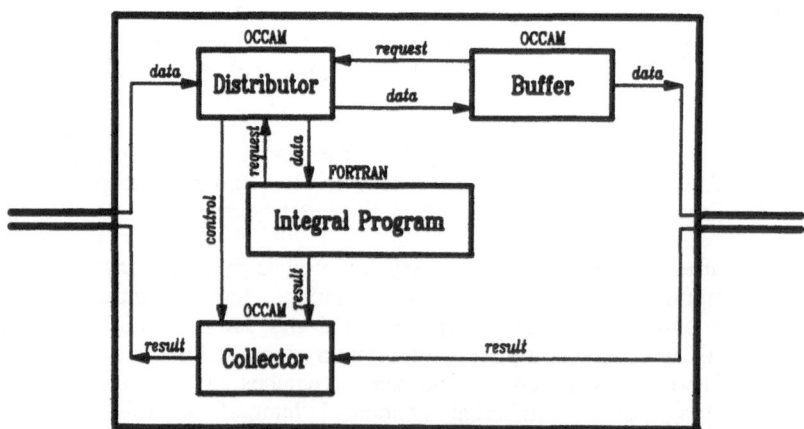

Figure 4. Processes, running on each network (integral) node.

In the present version of our direct SCF program, only this step is parallelized. All the other parts are executed sequentially on the master node. When in the course of the execution the two electron integrals are required, the master node sends via OCCAM channels quadruples of shell indices to the integral nodes (worker) in the network and processes the incoming integrals quasiparallelly to build the two electron part of the Fock matrix. The processes running on each of the integral nodes are illustrated in Figure 4. The largest process consists of code taken from the conventional program package MELD [24]. The I/O parts in the code were substituted by the subroutines CHANINMESSAGE and CHANOUTMESSAGE (Parsytec Ltd.) to enable the communication with the OCCAM processes via channels.

The Distributor process sends incoming data either to the Integral Program or to the next node or in both directions, depending on the state of the other processes and on the type of the data. Subsequently its most important OCCAM statements are summarized.

```
PROC distributor ( parameter ... )
    ... declarations
    WHILE TRUE
        SEQ
            ... read data
            IF
                (data.flag = ...)  --  data for all nodes
                    PAR
                        ... send data to the next node
                        ... send data to the integral program
                (data.flag = ...)  --  data for one batch of integrals
                    PRI ALT
                        request.buffer ? free
                        ... send data to the next node (buffer)
                        request.integral.program ? free
                        ... send data to the integral program
            TRUE
                SKIP
    :
```

All data packets are preceded by a flag which indicates a broadcast message or individual data. The latter are distributed using the ALT construct. ALT is comparable to the IF statement in other languages, except the fact that the status of input channels, in our case the request channels, is used as criterion for the branching and not the value of a logical value or expression. An input channel is ready, if the process on the other side wants to send. The *Integral Program* sends a signal via the request channel at the start of the program and every time the calculation of one batch has finished. The *Buffer* sends a request signal at the beginning and after each transfer of a data packet to the next node.

Control characters, geometrical data of the molecule and the basis set information are broadcasted to all integral nodes. If we wouldn't store this information on every node, but send, for each batch, the necessary data together with the quadruple of shell indices, extensive communication would decrease the performance significantly.

With the present version of the processes we can configure linear farms. Three of them may be connected to the master node. A tree structure is possible after little modifications. Therefore the *Buffer* process is included in order to save external links on the nodes. The *Collector* is a simple multiplexer process (ALT).

The whole distributed program runs either under the control of MultiTool or as a stand alone program communicating with the server process on the host computer.

TESTING THE DIRECT SCF PROGRAM

First test calculations were carried out on a network, consisting of nine transputers T800 in the 20 MHz version. One of the nodes, the master (SCF) node, was equipped with 4 MB of local memory. The other eight (integral) nodes each had 1 MB RAM. We used an older version of MultiTool, called MEGATOOL 4.2 as development system.

The details of the calculations are explained in ref. [4]. We investigated several network configurations differing in the number of processors and the number of farms attached to the master node. Some of the measured execution times for the parallelized step as well as for a whole iteration are shown in Table 1. From all calculations we selected the best (benzene with a basis of 78 s- and p-functions (DZ)) and the worst (formic acid with a basis of 58 s-, p- and d-functions (DZP)) case with respect to the increase of performance relative to the number of processors.

Table 1. Execution time (in seconds) measured for the calculation of two electron integrals and for the first SCF iteration. The speed-up factors relative to configuration I are given in brackets.

configuration	I		II		III		IV	
integral nodes	1		8		8		8	
farms	1 [1]		1 [8]		2 [4/4]		3 [3/3/2]	
	calculation of the two electron integrals							
C_6H_6 DZ-Basis	19 962	(1.0)	2 549	(7.8)	2 518	(7.9)	2 509	(8.0)
$HCOOH$ DZP-Basis	2 583	(1.0)	334	(7.7)	328	(7.9)	327	(7.9)
	first SCF iteration							
C_6H_6 DZ-Basis	20 110	(1.0)	2 592	(7.8)	2 568	(7.8)	2 563	(7.8)
$HCOOH$ DZP-Basis	2 672	(1.0)	373	(7.2)	373	(7.2)	374	(7.1)

Table 2. Calculation of the two electron integrals for trans formic acid (basis: 58 s-, p- and d-functions).

	computer	program	time	
a)	CYBER 175	ATMOL	524	c)
a)	CYBER 855	ATMOL	387	c)
a)	CYBER 205	ATMOL	191 – 200	c)
a)	CRAY-1S	ATMOL	132	c)
a)	CYBER 205	HONDO/COLUMBUS	88 – 93	c)
b)	IBM-4383	GAMESS	338	c)
b)	FPS-164/MAX-3	GAMESS	320	c)
b)	LCAP: 4 prozesses	GAMESS	137	c)
	MicroVAX II	MELD	1948	c)
	Comparex 7/78	MELD	125	c)
	CRAY-X/MP (1 prozessor)	MELD	76	c)
	configuration I (1 integral node)		2583	d)
	configuration IV (8 integral nodes, 3 farms)		327	d)

a)   see ref. [3] and references therein.

b)   see ref. [10]. The CPU time given for the LCAP is the result of a simulation.

c)   CPU times in seconds.

d)   execution times in seconds.

Configuration II, where we use only one linear farm, is a little more unfavorable than configuration III and IV, where the integral nodes are split into several farms. This is due to the fact that more communication has to be done per node in configuration II. However the diminution of performance is so small that we do not consider the communication being a problem even after an extension of the system.

The load balance of the network amounts to practically 100 % in the integral calculation step. This is made possible by the farming mechanism, despite the different size of the integral batches. Predetermining the node where a specific batch has to be calculated would lead to a worse load balance in many cases [9,10].

Only a small fraction of the total execution time is needed for the sequential parts of the program running on the master node. Therefore we observed a total speed-up of at least 7.1 for the first SCF iteration. With an increasing number of integral nodes however, the sequential part will be of more consequence and effectively parallelized algorithms for matrix operations are more necessary. In our group corresponding investigations are in progress.

In Table 2 we compare the results of our work with CPU times observed with different conventional SCF programs on a variety of computers. A negative as well as a positive aspect can be derived from the data. The former is the fact that the execution time with 1 integral node (configuration I) is higher than the CPU time measured with MELD on a MicroVAX II, although nearly the same subroutines are used for the calculations, and standard benchmarks show, that the T800 should be at least twice as fast as the VAX processor. Our FORTRAN compiler has no optimizing options and is surely less sophisticated than the VAX/VMS compiler. But that is not the only reason. In the MELD integral program, many intermediate results with respect to a product of basis functions (*charge distribution*) are calculated once and stored on mass storage devices for multiple use. As we do not use global memory and the local memory is too small, we have to recalculate the data every time we need them. That is the price that we have to pay to be able to distribute the program on a local memory parallel system.

However the price is worthwhile as the performance increases nearly linear relative to the number of processors. With configuration IV we get 1/3 to 1/2 of the performance of a fast scalar mainframe like the Comparex 7/78. A CRAY-X/MP processor is only 3 to 5 times faster, depending of the basis used. By the way the numbers in Table 2 again indicate the modest vectorizability of standard SCF code.

IMPROVING THE PERFORMANCE

The numbers discussed in the last sections are the results obtained with the first working version of our direct SCF program. To further reduce the execution time we perform investigations on three levels, the hardware, the algorithmic and the software level.

More and faster nodes

Recently we extended our system by four nodes being equipped with the 25 MHz T800 transputer. We configured one linear farm consisting of nine 20 MHz and three 25 MHz integral nodes being attached to a 25 MHz master (SCF) node. Taking into account the number and the different speed (factor 1.25 for the 25 MHz version) of the integral nodes a maximum speed-up of 1.59 can be gained compared to configuration II with one linear farm of eight 20 MHz nodes.

The observed speed-up amounted to 1.51 for the calculation of the two electron integrals of trans formic acid (DZP basis, cf. Tables 1 and 2). 1.56 was measured when calculating P_4S_3 with a STO-3G basis. The speed-up increases nearly linear as predicted, and our farming mechanism really can balance the load of a network consisting of nodes with differing performance.

The adaption of the program to the new network was very easy. We only had to change six statements and recompile and reconfigure the code.

Together with the 25MHz master node we installed the new version of MultiTool [16]. This software besides other improvements now contains an OCCAM compiler that allows the use of functions. Due to this fact as well as the faster hardware the sequential part of the program runs twice as fast as the former version. The integral processes written in FORTRAN are not affected by the new MultiTool.

Faster algorithms for the calculation of the two electron integrals

In the calculations discussed up to now we used the McMurchie-Davidson algorithm [25] to evaluate the integrals. Especially for basis functions with low angular momentum faster algorithms are available. One of these was developed by Bär [26] based on the work of Obara and Saika [27] and is used in the TURBOMOLE program package [23] with success. We now integrate step by step the subroutines worked out by Bär in our integral program. The benefits of the new routines can be seen in Table 3.

Table 3. Calculation of the two electron integrals using the McMurchie-Davidson algorithm (MD) [25] and some subroutines developed by Bär (B) [26]: Execution time in seconds.

B		-	$(ss\lvert ss)$	$(ss\lvert ss)$ $(ps\lvert ss)$	$(ss\lvert ss)$ $(ps\lvert ss)$ $(pp\lvert ss)$ $(ps\lvert ps)$
MD		all	rest	rest	rest
C_2H_4	DZ	(a)	92	77	46	27
C_2H_4	DZPD	(b)	487	465	391	302
$HCOOH$	DZP	(c)	334	325	295	252

a) 32 s- and p-functions
b) 64 s-, p- and d-functions
c) 58 s-, p- and d-functions

Better utilization of intermediate results

In the first version of the program we calculated one batch of integrals $(ij\lvert kl)$ with a given quadruple of shell indices i, j, k and l independently from all the other batches. The decrease of performance due to the repeated calculation of intermediate results was discussed above. We expected a smaller effect when evaluating more than one batch in a single process. For example we collected the batches being derived from a fixed charge distribution ij and all charge distributions kl with a canonical index $I_{kl} = k(k-1)/2 + l$ less or equal I_{ij} in one group.

In fact the execution time was halved in smaller networks. If the number of integral nodes increases however we run into load balancing problems even with the farming mechanism. The difference in the granularity of the independent processes is by some orders of magnitude larger as in the case where we treat single batches. We have to investigate other ways of partitioning the two electron integrals into larger groups of batches.

FIRST EXPERIENCE WITH HELIOS

Under Helios a farm like our direct SCF program can easily be defined. The farm constructor

master(SCF) [n] ||| *worker(integrals)*

is an element of the Component Distribution Language (CDL). *n* is the number of workers that will communicate bidirectionally with the master. The data flow to the workers is controlled by a load balancer process (lb) which is supplied with Helios. The farm constructor is an abbreviation of the following statement.

master <> *lb n* (<> *worker*(1), <> *worker*(2),..., <> *worker*(n))

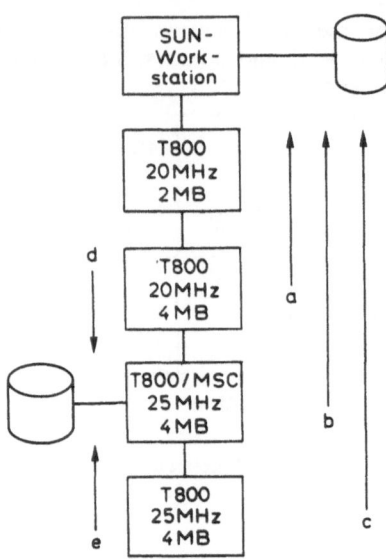

Figure 5. Configuration of the 4 MB nodes in our Helios network. The data paths a - e are described in Table 4.

The user has not to care for the configuration of the network and for the distribution of the tasks. This is done by the Task Force Manager.

The master and the worker tasks may be standard FORTRAN programs using Fortran I/O statements. Unfortunately Helios is a new piece of software and like all other new software it contains a lot of bugs. One of them concerns the interface between the FORTRAN compiler and CDL. Further comments on the Fortran compiler can be found in ref. [28]. Up to now we couldn't implement our distributed direct SCF code under Helios.

Experience could be gained however with the distribution of whole programs in the network. One of these programs is the semidirect SCF part of the TURBOMOLE package [23]. Its more than 15 000 lines of FORTRAN code were compiled with no modifications. We succeeded after using the -Zn option which allocated more memory to the compiler. This option is not documented in the manuals.

Another problem arose from the wrong execution of the status option in the FORTRAN OPEN and CLOSE statements. Besides this we could run the program without difficulties.

As the transputer has no virtual memory management, the whole executable program has to fit into the physical memory of a node. The DSCF module needs about 3 MB even with relative small array dimensions. Therefore we could only use the 4 MB nodes in our network (Figure 5).

The TURBOMOLE DSCF is a semidirect SCF program. The most expensive integrals are stored on disk and only the rest is recalculated in every iteration. We can access via links either the disks of the SUN workstation (a - c in Figure 5) or the disk attached to the SCSI interface of the MSC node (d and e in Figure 5). In the latter case we cannot use the MSC node for the calculations as the file server occupies 2.7 MB of local memory.

Table 4. Total execution times (in s) for the calculation of formic acid with a DZP basis (55 s-, p- and d-functions).

node / path	single process	a + b parallel	a + b + c parallel	d + e parallel
a	3880	3930	4010	
b	3130	3170	3290	
c	3190		3250	
d	4149			4229
e	3413			3655

a) T800(20 MHz)   - via 2 links   - SUN disk
b) T800(25 MHz)   - via 3 links   - SUN disk
c) T800(25 MHz)   - via 4 links   - SUN disk
d) T800(20 MHz)   - via 1 link   - MSC SCSI disk
e) T800(25 MHz)   - via 1 link   - MSC SCSI disk

In our test job we calculated trans formic acid with the DZP basis from the TURBOMOLE library. The size of the integral file amounted to about 6 MB. The execution times for the complete calculation are summarized in Table 4. The times in the first column were measured with no other large processes running in parallel. The T800/MSC 25 MHz node (b) is 1.24 times as fast as the 20 MHz node. Practically no difference can be seen if we route the integral stream either through one (a) or through two (b) additional nodes. Even in case c with three intermediate nodes the increase of the execution time is acceptable.

The combination MSC - SCSI interface - disk together with the Helios file system [18] used in d) and e) is slower than the access to the SUN disk. This effect is even larger if several SCF processes are executed parallel in the network (compare column 2 and 4 in Table 4).

At least up to three processors we observe a nearly linear increase of the total performance of the network, relative to the number of nodes. The communication via links and their control by the transputer itself is efficient. The execution of the SCF program is not disturbed significantly by the fact that other processes route a lot of data through the node.

CONCLUSION

The direct SCF procedure is very suitable for the implementation on massively parallel computers. The by far most time consuming step, the calculation of the two electron integrals can be parallelized with an excellent load balance. Even if all the other parts of the program run sequentially on a single node, the performance increases nearly linear with the number of processors. This is true at least for the 12 integral nodes of our transputer network. We expect a further increase if we expand the system.

Farming on the subroutine level proved to be a useful concept to guarantee a good load balance. Its second advantage is the fact that we can easily integrate existing code in the arming harness. Thus the time to develop a program can be reduced considerably. In addition we can participate in the successful investigations of other groups which nearly exclusively program their improved algorithms in FORTRAN. As an example we inserted some of the TURBOMOLE integral subroutines [23,26].

The communication processes necessary to build up a farm were developed in our group and programmed in OCCAM. In future this work may be saved by using distributed operating systems like Helios. They promise an even more easy way to transfer standard FORTRAN code to parallel systems. In a first step we executed whole programs (TURBOMOLE [23]) in parallel. The use of only one disk for several tasks turned out to be no bottleneck. This demonstrates the very good suitability of the transputer for message passing systems. The only limitation for the use of whole programs seems to be the size of the local memory of a node.

Parallel computers with distributed local memory can be used in quantum chemistry very efficiently. The transputer was designed for parallel systems and computing nodes can be constructed with little expense. Transputer systems offer a low price/performance ratio. They are a favorable tool to gain experience with parallel computing.

REFERENCES

1. V.R. Saunders and M.F. Guest, *Comp. Phys. Comm.* 26:389-395 (1982).
2. R. Ahlrichs, H.-J. Böhm, C. Ehrhardt, P. Scharf, H. Schiffer, H. Lischka and M. Schindler, *Journ. of Comp. Chem.* 6:200-208 (1985).
3. W. Kutzelnigg, M. Schindler, W. Klopper, S. Koch, U. Meier and H. Wallmeier, in *Supercomputer Simulations in Chemistry*, M. Dupuis, ed., Springer, Berlin (1986), p. 55-74 (Lecture notes in chemistry 44).
4. U. Wedig, A. Burkhardt, and H.-G. von Schnering, *Z. Phys. D - Atoms, Molecules and Clusters* 13:377-384 (1989).
5. M.E. Colvin, Ph. D. thesis LBL--23578, Lawrence Berkeley Laboratory, University of California, Berkeley, California 94720 (1986).
6. R.A. Whiteside, J.S. Binkley, M.E. Colvin, and H.F. Schaefer III, *Journ. Chem. Phys.* 86:2185-2193 (1987).
7. E. Clementi, *Philos. Trans. R. Soc. London* A326:445-470 (1988).
8. E. Clementi, G. Corongiu, J. Detrich, S. Chin and L. Domingo, *Int. Journ. Quantum Chem.: Quantum Chem. Symp.* 18:601-618 (1984).
9. M. Dupuis and J.D. Watts, *Theor. Chim. Acta* 71:91-103 (1987).
10. M.F. Guest, R.J. Harrison, J.H. van Lenthe and L.C.H. van Corler, *Theor. Chim. Acta* 71:117-148 (1987).
11. D.A. Buell, et al., *The Journal of Supercomputing* 1:301-325 (1988).
12. INMOS Ltd., *The Transputer Family* 1987, INMOS Product Information, INMOS, Bristol (1987).
13. INMOS Ltd., *OCCAM 2 Reference Manual*, Prentice Hall, London (1988). (Prentice Hall International Series in Computer Science).
14. A.J.G. Hey, *Comp. Phys. Comm.* 50:23-31 (1988).
15. A.J.G. Hey, *Comp. Phys. Comm.* 56:1-24 (1989).
16. *MultiTool 5.0*, Parsytec Ltd., Aachen.
17. Perihelion Software Ltd., *The Helios Operating System*, Prentice Hall, London (1989).

18. *Helios file system*, version 1.1, Parsytec Ltd., Aachen.
19. I. Glendinning and A. Hey, *Comp. Phys. Comm.* 45:367-371 (1987).
20. J. Almlöf, K. Faegri Jr. and K. Korsell, *Journ. of Comp. Chem.* 3:385-399 (1982).
21. J. Almlöf, and P.R. Taylor, in *Advanced Theories and Computational Approaches to the Electronic Structure of Molecules*, C.E. Dykstra, ed., Reidel, Dordrecht, (1984), 107-125. (NATO ASI series C 133).
22. D. Cremer and J. Gauss, *Journ. of Comp. Chem.* 7:274-282 (1986).
23. M. Häser, and R. Ahlrichs, *Journ. of Comp. Chem.* 10:104-111 (1989).
24. L. McMurchie, S.T. Elbert, S.R. Langhoff, E.R. Davidson, et al., *Program MELD* (University of Washington, Seattle). Modified version: U. Wedig.
25. L.E. McMurchie and E.R. Davidson, *Journ. of Comp. Phys.* 26:218 (1978).
26. M. Bär, Diploma Thesis, University of Karlsruhe (1988).
27. S. Obara and A. Saika, *Journ. Chem. Phys.* 84:3963-3974 (1986).
28. R. Allan, *Parallelogram* 2(21):17 (1989).

DOMAIN DECOMPOSITION METHODS FOR PARTIAL DIFFERENTIAL

EQUATIONS AND PARALLEL COMPUTING

François-Xavier Roux

O.N.E.R.A., 29 Av. de la Division Leclerc, BP 72
F-92322 CHATILLON Cedex, France

ABSTRACT

We present the most classical domain decomposition methods for solving elliptic partial differential equations. The first one, the Schwarz alternative procedure, involves overlapping subregions, the two others involve non-overlapping subdomains: a conforming method, the Schur complement method, and a non-conforming method, based on introducing a Lagrange multiplier in order to enforce the continuity requirements at the interface between the subdomains.

All these methods are analyzed in terms of condensed problems on the interface.

The problem of the parallel implementation of these methods is addressed, and the results of some numerical experiments for ill conditioned three-dimensional structural analysis problems are given.

1. INTRODUCTION

The simplest parallel algorithm for solving elliptic partial differential equations, with finite element or finite difference discretization, is based on solving the complete problem through the conjugate gradient method with parallelization of the matrix-vector product. This can be done by performing in parallel the computation for the lines or the rows of the matrix associated with different substructures. But, with sparse matrices arising from finite element methods, the amount of computation of a matrix-vector product depends in a linear way on the number of variables. As the data transfers depend in a linear way on the number of variables too, the parallelization of these problems typically leads to fine grain parallelism.

Moreover, for large structural analysis problems, using a global conjugate gradient method is really problematic, because of the ill conditioning and the large numbers of degrees of freedom.

It is possible to decrease the dimension of the problem and to get parallel algorithms with large granularity by using domain decomposition methods. Some of these methods involve overlapping subregions and are derived from the Schwarz alternative principle.

Scientific Computing on Supercomputers II
Edited by J. T. Devreese and P. E. Van Camp
Plenum Press, New York, 1990

139

Other methods involve non-overlapping subregions. These methods consist in solving a condensed problem on the interface between the subdomains. The condensed operator is defined with the help of the inverses of local matrices associated with independent local problems. These methods appear to be better suited to finite element or finite difference methods, first, because they lead to solve the condensed problem on the interface through the preconditioned conjugate gradient method, and secondly, because it is generally more difficult to split an unstructured mesh in overlapping than in non-overlapping subregions. And, at last, the Schwarz alternative method is less intrinsically parallel.

In this paper, we present, first, the standard Schwarz alternative procedure, secondly, the most classical domain decomposition method with non-overlapping subregions: the Schur complement method. Thirdly, we present a non-conforming method, based on introducing a Lagrange multiplier to enforce the continuity requirement at the interface between the subdomains, that we call the hybrid method, because it is similar to the well-known hybrid finite element method for the elasticity equation. We show that the two methods with non-overlapping subdomains are associated with dual formulation of the condensed problem on the interface. Then, we address some practical problems with the implementation of these methods, concerning the topology of the interface, and the choice of the local solver. At last, we give some results with the implementation of the hybrid method for solving an ill-conditioned three-dimensional structural analysis problem.

In the sequel of this paper we shall use the vocabulary of the linear elasticity. But the methods presented here can, of course, apply to any second order elliptic partial differential equations.

2. THE SCHWARZ ALTERNATIVE PRINCIPLE

2.1. Presentation of the Method

Note $\mathbf{Lu} = \mathbf{f}$, an elliptic partial differential equation to be solved on a domain Ω. Consider a splitting of the domain into three distinct open regions Ω_1, Ω_2 and Ω_3, and their internal interfaces Γ_4 and Γ_5, as presented in Figure 1.

So, the domain can be decomposed into two overlapping open subregions: $\Omega_{12} = \Omega_1$ U Γ_4 U Ω_2 and $\Omega_{23} = \Omega_2$ U Γ_5 U Ω_3.

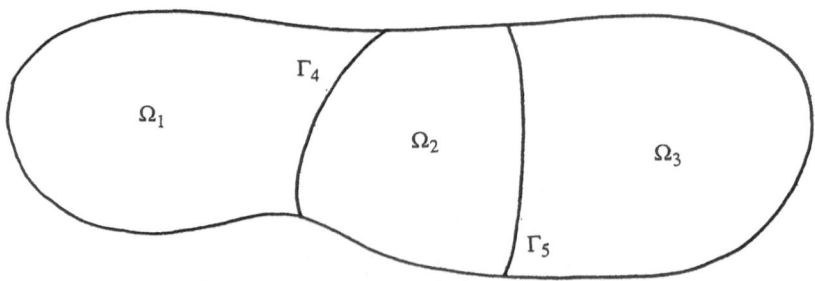

Figure 1. Overlapping domain decomposition.

The Schwarz alternative procedure consists in iteratively solving the problem in each subregion, with boundary conditions on the internal artificial interface defined as the restriction of the solution in the other region at the previous iteration.

The iteration number of the Schwarz alternative method can be written as follows:

given $\mathbf{u}_5^n$ on Γ_5, solve the problem $\mathbf{Lu} = \mathbf{f}$ on Ω_{12} with the Dirichlet boundary conditions: $\mathbf{u} = \mathbf{u}_5^n$ on Γ_5

take the restriction of the solution $\mathbf{u}_{12}^{n+1/2}$ on Γ_4: $\mathbf{u}_4^{n+1/2}$

solve the problem $\mathbf{Lu} = \mathbf{f}$ on Ω_{23} with the Dirichlet boundary conditions: $\mathbf{u} = \mathbf{u}_4^{n+1/2}$ on Γ_4,

take the restriction of the solution $\mathbf{u}_{23}^{n+1}$ on Γ_5: $\mathbf{u}_5^{n+1}$,

If $\mathbf{u}_5^{n+1} = \mathbf{u}_5^n$, then the local solutions $\mathbf{u}_{12}^{n+1/2}$ and $\mathbf{u}_{23}^{n+1}$ are equal to the restrictions to Ω_{12} and Ω_{23} of the solution of the complete problem.

The procedure can be proved to converge for various geometries of the interface, with a convergence ratio depending upon the size of the overlapping subregion (see [1]).

2.2. Formulation of the Method in Terms of the Interface Operator

By ordering the unknowns corresponding to Ω_1, Ω_2 and Ω_3 before those associated with the degrees of freedom located on Γ_4 and Γ_5, the stiffness matrix of the problem has the following block structure:

$$
\mathbf{K} = \begin{bmatrix}
\mathbf{K}_{11} & 0 & 0 & \mathbf{K}_{14} & 0 \\
0 & \mathbf{K}_{22} & 0 & \mathbf{K}_{24} & \mathbf{K}_{25} \\
0 & 0 & \mathbf{K}_{33} & 0 & \mathbf{K}_{35} \\
\mathbf{K}_{41} & \mathbf{K}_{42} & 0 & \mathbf{K}_{44} & 0 \\
0 & \mathbf{K}_{52} & \mathbf{K}_{53} & 0 & \mathbf{K}_{55}
\end{bmatrix}
$$

As the variables associated with the different open subsets Ω_1, Ω_2 and Ω_3 are decoupled, the global problem $\mathbf{Ku} = \mathbf{b}$ can be replaced, by a block-Gaussian elimination of $\mathbf{u}_1$, $\mathbf{u}_2$ and $\mathbf{u}_3$, by the following condensed problem on the interfaces Γ_4 and Γ_5

$$
\begin{bmatrix}
\mathbf{M}_4 & -\mathbf{K}_{42}\mathbf{K}_{22}^{-1}\mathbf{K}_{25} \\
-\mathbf{K}_{52}\mathbf{K}_{22}^{-1}\mathbf{K}_{24} & \mathbf{M}_5
\end{bmatrix}
\begin{bmatrix}
\mathbf{u}_4 \\
\mathbf{u}_5
\end{bmatrix}
=
\begin{bmatrix}
\mathbf{g}_4 \\
\mathbf{g}_5
\end{bmatrix}
, \qquad (1)
$$

with

$$
\mathbf{M}_4 = \mathbf{K}_{44} - \mathbf{K}_{41}\mathbf{K}_{11}^{-1}\mathbf{K}_{14} - \mathbf{K}_{42}\mathbf{K}_{22}^{-1}\mathbf{K}_{24} ,
$$

$$
\mathbf{M}_5 = \mathbf{K}_{55} - \mathbf{K}_{51}\mathbf{K}_{11}^{-1}\mathbf{K}_{15} - \mathbf{K}_{52}\mathbf{K}_{22}^{-1}\mathbf{K}_{25} ,
$$

$$\mathbf{g}_4 = \mathbf{b}_4 - \mathbf{K}_{41}\mathbf{K}_{11}^{-1}\mathbf{b}_1 - \mathbf{K}_{42}\mathbf{K}_{22}^{-1}\mathbf{b}_2 \ ,$$

$$\mathbf{g}_5 = \mathbf{b}_5 - \mathbf{K}_{51}\mathbf{K}_{11}^{-1}\mathbf{b}_1 - \mathbf{K}_{52}\mathbf{K}_{22}^{-1}\mathbf{b}_2 \ ,$$

Let us now consider the iteration number n+1 of the Schwarz algorithm. The first step consists in computing the solution of the problem in Ω_{12} with the Dirichlet boundary conditions: $\mathbf{u} = \mathbf{u}_5^n$ on Γ_5, that satisfy the following linear system:

$$\begin{bmatrix} \mathbf{K}_{11} & 0 & \mathbf{K}_{14} \\ 0 & \mathbf{K}_{22} & \mathbf{K}_{24} \\ \mathbf{K}_{41} & \mathbf{K}_{42} & \mathbf{K}_{44} \end{bmatrix} \begin{bmatrix} \mathbf{u}_1 \\ \mathbf{u}_2 \\ \mathbf{u}_4 \end{bmatrix} = \begin{bmatrix} \mathbf{b}_1 \\ \mathbf{b}_2 - \mathbf{K}_{25}\mathbf{u}_5^n \\ \mathbf{b}_4 \end{bmatrix}$$

By elimination of the unknowns $\mathbf{u}_1$ and $\mathbf{u}_2$ in this system, one can see that $\mathbf{u}_4^{n+1/2}$, the restriction Γ_4 of the solution of the problem above, satisfies the following condensed problem:

$$\mathbf{M}_4 = \mathbf{u}_4^{n+1/2} = \mathbf{b}_4 - \mathbf{K}_{42}\mathbf{K}_{22}^{-1}\mathbf{K}_{25}\mathbf{u}_5^n$$

The same analysis shows that the second step of the Schwarz alternative procedure consists in solving the condensed problem on Γ_5

$$\mathbf{M}_5 = \mathbf{u}_5^{n+1/2} = \mathbf{b}_5 - \mathbf{K}_{42}\mathbf{K}_{22}^{-1}\mathbf{K}_{25}\mathbf{u}_4^{n+1/2}$$

This analysis which was first developed in [2], proves that the Schwarz alternative procedure is precisely a block Gauss-Seidel algorithm for solving the condensed interface problem (1).

2.3. Parallel Implementation of the Schwarz Alternative Procedure

The Schwarz procedure is not intrinsically parallel, because the computation of the solution in one subdomain depends upon the solution in the other subdomain. So, it is not possible to compute in parallel the different local solutions.

To build a parallel algorithm with the Schwarz alternative procedure, the splitting of the domain must be done in such a way that each one of the two subregions Ω_{12} and

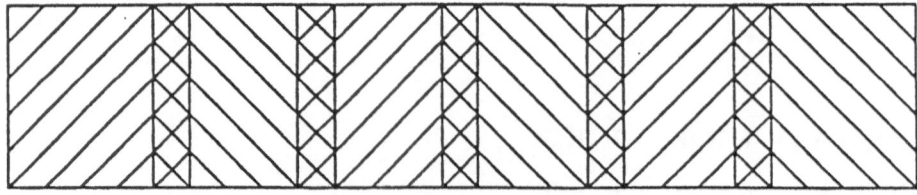

Figure 2. Red-black decomposition.

Ω_{23} is made of disjoined connected open subsets. The simplest example of such a splitting consist in cutting a domain into overlapping slices with a red-black coloring of the subsets. Then, the domain Ω_{12} consists of the red sections and the domain Ω_{23} of the black ones.

So, the computation of the solution in the domain Ω_{12} involves the of independent problems in the different red sections that can be performed in parallel.

2.4. Some Remarks about the Schwarz algorithm

The Schwarz alternative algorithm is not intrinsically parallel, and leads to solve the problem twice at each iteration on the overlapping regions of the domain. The red-black decomposition of the domain allows to work in parallel in subsets covering at most half the global domain. Furthermore, splitting an unstructured mesh into overlapping subregions may be more difficult than to decompose it into non-overlapping subdomains. So the method is not very well suited for finite element problems. From a linear algebra viewpoint, using a block Gauss-Seidel method for solving a symmetric problem is generally not an optimal solution.

So, the Schwarz method is, a priori, less efficient than the domain decomposition methods with non-overlapping subdomains that will be presented in the next sections of this paper. Nevertheless, recent research work has been done in order to find more parallel versions of this algorithm, and to accelerate the convergence, for instance by mixing Dirichlet and Neumann boundary conditions on the internal boundaries (see [3]). At last, the method can be effective for special geometries, in the case where the domain can be naturally split into overlapping substructures with such a shape that fast local solvers can be used.

3. THE SCHUR COMPLEMENT METHOD

3.1. Presentation of the Method

The most classical decomposition method with non-overlapping subdomains, the so-called Schur complement method, is based on the Gaussian elimination of degrees of

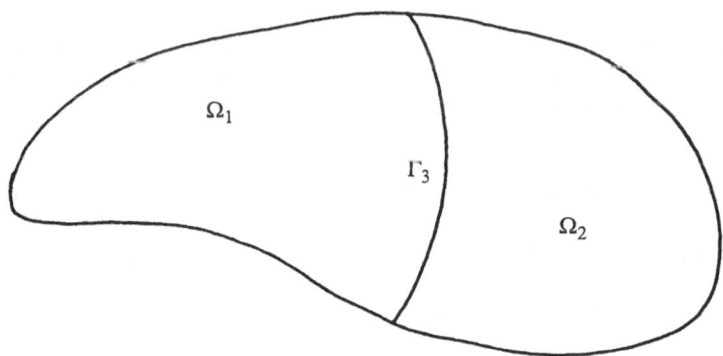

Figure 3. Non-overlapping domain decomposition.

freedom inside the substructures. Consider the linear elasticity equations on a domain Ω, and $\mathbf{K}$ the matrix associated with a Lagrangian finite element approximation of the displacement fields. Split the domain Ω into two open subsets Ω_1 and Ω_2 with Γ_3 the inner intersection of the boundaries Γ_1 and Γ_2 of Ω_1 and Ω_2.

The stiffness matrix associated with the renumbering of the degrees of freedom according to the splitting of the domain into these three subsets can be written in the block form below:

$$
\mathbf{K} = \begin{bmatrix} \mathbf{K}_{11} & 0 & \mathbf{K}_{13} \\ 0 & \mathbf{K}_{22} & \mathbf{K}_{23} \\ \mathbf{K}^{13} & \mathbf{K}^{23} & \mathbf{K}_{33} \end{bmatrix}
$$

The stiffness matrix associated with the linear elasticity equations on Ω_1 and Ω_2 with Neumann boundary conditions on Γ_3 are:

$$
\mathbf{K}^{(1)} = \begin{bmatrix} \mathbf{K}_{11} & \mathbf{K}_{13} \\ \mathbf{K}^{13} & \mathbf{K}_{33}^{(1)} \end{bmatrix} \quad \text{and } \mathbf{K}^{(2)} = \begin{bmatrix} \mathbf{K}_{22} & \mathbf{K}_{23} \\ \mathbf{K}_{23} & \mathbf{K}_{33}^{(2)} \end{bmatrix}
$$

The coefficients of the matrices $\mathbf{K}_{33}^{(1)}$ and $\mathbf{K}_{33}^{(2)}$ are the contributions from the integrals over Ω_1 and Ω_2 of the basis functions associated with the nodes of Γ_3, and so $\mathbf{K}_{33} = \mathbf{K}_{33}^{(1)} + \mathbf{K}_{33}^{(2)}$.

To solve the global problem,

$$
\mathbf{K}\,\mathbf{x} = \mathbf{b} \quad \text{with } \mathbf{x} = \begin{bmatrix} \mathbf{x}_1 \\ \mathbf{x}_2 \\ \mathbf{x}_3 \end{bmatrix} \quad \text{and } \mathbf{b} = \begin{bmatrix} \mathbf{b}_1 \\ \mathbf{b}_2 \\ \mathbf{b}_3 \end{bmatrix}
$$

one can perform a Gaussian elimination of the degrees of freedom inside the open subsets Ω_1 and Ω_2, and get the following condensed problem, involving only the degrees of freedom on Γ_3:

$$
\left[\mathbf{K}_{33} - \mathbf{K}_{13}^{t}\,\mathbf{K}_{11}^{-1}\,\mathbf{K}_{13} - \mathbf{K}_{23}^{t}\,\mathbf{K}_{22}^{-1}\,\mathbf{K}_{23} \right] \mathbf{x}_3 = \mathbf{b}_3 - \mathbf{K}_{13}^{t}\,\mathbf{K}_{11}^{-1}\,\mathbf{b}_1 - \mathbf{K}_{23}^{t}\,\mathbf{K}_{22}^{-1}\,\mathbf{b}_2 \tag{2}
$$

The associated matrix, the Schur complement matrix $\mathbf{S}$, is symmetric and positive definite (see [4]). This is proved by the following inequality that consists in a change of basis for the dot product associated with the $\mathbf{K}$ matrix:

$$
\begin{bmatrix} \mathbf{I} & 0 & 0 \\ 0 & \mathbf{I} & 0 \\ -\mathbf{K}_{31}\mathbf{K}_{11}^{-1} & -\mathbf{K}_{32}\mathbf{K}_{22}^{-1} & \mathbf{I} \end{bmatrix} \begin{bmatrix} \mathbf{K}_{11} & 0 & \mathbf{K}_{13} \\ 0 & \mathbf{K}_{22} & \mathbf{K}_{23} \\ \mathbf{K}_{31} & \mathbf{K}_{32} & \mathbf{K}_{33} \end{bmatrix} \begin{bmatrix} \mathbf{I} & 0 & -\mathbf{K}_{11}^{-1}\mathbf{K}_{13} \\ 0 & \mathbf{I} & -\mathbf{K}_{22}^{-1}\mathbf{K}_{13} \\ 0 & 0 & \mathbf{I} \end{bmatrix} = \begin{bmatrix} \mathbf{K}_{11} & 0 & 0 \\ 0 & \mathbf{K}_{22} & 0 \\ 0 & 0 & \mathbf{S} \end{bmatrix}
$$

So, the problem (2) can be solved through the conjugate gradient method without actually computing the coefficients of the matrix $\mathbf{S}$.

Let $\mathbf{x}_3$ be a given displacement vector on $\Gamma 3$. The product $\mathbf{Sx}_3$ is given by:

$$\mathbf{Sx}_3 = \mathbf{S}^{(1)}\mathbf{x}_3 + \mathbf{S}^{(2)}\mathbf{x}_3$$

with

$$\mathbf{S}^{(1)} = \mathbf{K}_{33}^{(1)} - \mathbf{K}^{13} \mathbf{K}_{11}^{-1} \mathbf{K}_{13} \quad \text{and} \quad \mathbf{S}^{(2)} = \mathbf{K}_{33}^{(2)} - \mathbf{K}^{23} \mathbf{K}_{22}^{-1} \mathbf{K}_2$$

Computing the product $\mathbf{S}^{(1)}\mathbf{x}_3$ involves two steps. The first step is the solution of the Dirichlet problem on Ω with the boundary conditions given by $\mathbf{x}_3$ on Γ_3:

$$\mathbf{K}_{11} \mathbf{x}_1 = -\mathbf{K}_{13} \mathbf{x}_3$$

The second step is the computation of the product of the $\mathbf{K}^{(1)}$ matrix by the vector $(\mathbf{x}_1, \mathbf{x}_3)^t$ that is easily shown to be equal to $(0, \mathbf{S}^{(1)}\mathbf{u}_3)^t$.

So, solving the equation (2) through the conjugate gradient method gives a parallel algorithm with a very good granularity because the main part of the work consists in the computation of local independent contributions to the product by the Schur complement matrix, that involves mainly the solution of independent local problems.

3.2. A Preconditioner for the Schur Complement Method

In the previous section, we have seen that the product by the local Schur complement matrix can be computed by solving a problem with Dirichlet boundary conditions on the interface and then computing the trace of the corresponding internal forces. This leads to the following equation:

$$\begin{bmatrix} 0 \\ \mathbf{S}^{(1)}\mathbf{x}_3 \end{bmatrix} = \begin{bmatrix} \mathbf{K}_{11} & \mathbf{K}_{13} \\ \mathbf{K}_{31} & \mathbf{K}_{33}^{(1)} \end{bmatrix} \begin{bmatrix} \mathbf{K}_{11} & \mathbf{K}_{13} \\ 0 & \mathbf{I} \end{bmatrix}^{-1} \begin{bmatrix} 0 \\ \mathbf{x}_3 \end{bmatrix}$$

The Schur complement matrix is the discrete operator associated with the mapping of the trace of displacements fields on the interface onto the trace of the internal forces. This mapping is a so-called Steklov-Poincaré's operator (see [5]).

So, the inverse of the local Schur complement matrix can be computed by mapping the trace of the internal forces field onto the trace of displacements field on the interface. This leads to solving a local problem with Neumann boundary conditions. Let us note:

$$\begin{bmatrix} \mathbf{x}_1 \\ \mathbf{x}_3 \end{bmatrix} = \begin{bmatrix} \mathbf{K}_{11} & \mathbf{K}_{13} \\ 0 & \mathbf{I} \end{bmatrix}^{-1} \begin{bmatrix} 0 \\ \mathbf{x}_3 \end{bmatrix} ,$$

equation (3) shows that $\mathbf{x}_3$ is the restriction on $\Gamma 3$ of the solution of the Neumann problem:

$$\begin{bmatrix} \mathbf{K}_{11} & \mathbf{K}_{13} \\ \mathbf{K}_{31} & \mathbf{K}_{33}^{(1)} \end{bmatrix} \begin{bmatrix} \mathbf{x}_1 \\ \mathbf{x}_3 \end{bmatrix} = \begin{bmatrix} 0 \\ \mathbf{S}^{(1)}\,\mathbf{x}_3 \end{bmatrix}$$

An efficient preconditioner for the condensed problem associated with the Schur complement matrix can be built with the following shape:

$$\mathbf{M} = \mathbf{D}_1\,[\mathbf{S}^{(1)}]^{-1}\,\mathbf{D}_1^t + \mathbf{D}_2\,[\mathbf{S}^{(2)}]^{-1}\,\mathbf{D}_2^t$$

where the $\mathbf{D}_i$ matrices are weighing matrices such that $\mathbf{D}_1 + \mathbf{D}_2 = \mathbf{I}_{|\Gamma_3}$ (see, for instance, [6] and [7]).

As the Schur complement matrix is a mapping of the trace of displacement field on the interface onto the trace of the internal forces, the residual of the conjugate gradient algorithm is homogeneous to a forces field, whenever the problem is related to the displacements field. The preconditioner presented here consists in mapping the gradient vector back in the primal space associated with the displacements.

Computing this preconditioner leads to the same degree of parallelism as the plain algorithm, because it consists mainly in solving independent local Neumann problems and then assembling the local contributions on the interface.

4. THE HYBRID ELEMENT METHOD

4.1. Principle of the Hybrid Method

Another domain decomposition method is based on a mechanical approach and involves the introduction of a Lagrange multiplier on the interface to remove the continuity constraint. Let us note the equations of the linear elasticity equations with homogeneous boundary conditions in the following way:

$$\begin{cases} \mathbf{Au} = \mathbf{f} & \text{in } \Omega \\ \mathbf{u} = 0 & \text{on } \Gamma_0 \end{cases} \quad \text{with } (\mathbf{Au})_i = \frac{\partial(a_{ijkh}\,\varepsilon_{kh}(\mathbf{u}))}{\partial x_j} \tag{4}$$

The usual variational form of this problem consists in finding $\mathbf{u}$ in $(\mathbf{H}^1(\Omega))^3$ satisfying the boundary condition $\mathbf{u} = 0$ on Γ_0 which minimizes the energy functional:

$$\mathbf{I}(\mathbf{v}) = \frac{1}{2}\,\mathbf{a}(\mathbf{v},\mathbf{v}) - (\mathbf{f},\mathbf{v}) \quad \text{with } \mathbf{a}(\mathbf{u},\mathbf{v}) = \int_\Omega a_{ijkh}\,\varepsilon_{kh}(\mathbf{u})\,\varepsilon_{ij}(\mathbf{v})\,dx .$$

Let us consider the same splitting of the domain as in the previous section. For a sake of simplicity, let us assume that the boundary of the interface Γ_3 is embedded in Γ_0, the part of the boundary of Ω with homogeneous Dirichlet conditions. Then, the traces on Γ_3 of the displacement fields $\mathbf{u}$ satisfying the boundary condition $\mathbf{u} = 0$ on Γ_0 belong to the space $(\mathbf{H}_{00}^{1/2}(\Gamma_3))^3$.

Solving the linear elasticity equation consists in finding two functions u_1 and u_2, in the functional spaces V_1 and V_2 of the fields belonging to $(H^1(\Omega_1))^3$ and $(H^1(\Omega_2))^3$ that satisfy the boundary conditions on Γ_1 and Γ_2, which minimize the sum of the energies: $I(v) = I_1(v_1) + I_2(v_2)$, with the continuity constraint: $v_1 = v_2$ on Γ_3. The dual form of the continuity condition is:

$$(v_1 - v_2, \mu)_{\Gamma_3} = 0 \quad \text{for each } \mu \text{ in } [(H_{00}^{1/2}(\Gamma_3))^3]'$$

The primal hybrid variational principle is based on removing the intersubdomain continuity constraint by introducing a Lagrange multiplier (see for instance [8] or [9]). Under the assumption that the so-called Ladyzenskaia-Babuska-Brezzi condition is satisfied:

$$Sup \ (v_1 - v_2, \mu) \geq C \ |\mu|_{[(H_{00}^{1/2}(\Gamma_3))^3]'} \ ,$$

$$|(v_1 - v_2)|_{V_1 \times V_2} = 1$$

one can show (see [10]) that the problem of minimization with constraint above is equivalent to finding the saddle-point of the Lagrangian:

$$L(v, \mu) = I_1(v_1) + I_2(v_2) + (v_1 - v_2, \mu)_{\Gamma_3} . \tag{5}$$

This means finding the fields (u_1, u_2) in $V_1 \times V_2$ and the Lagrange multiplier λ in $[(H_{00}^{1/2}(\Gamma_3))^3]'$ which verify:

$$L(u, \mu) \leq L(u, \lambda) \ L(v, \lambda)$$

for each field $v = (v_1, v_2)$ in $V_1 \times V_2$, and each μ in $[(H_{00}^{1/2}(\Gamma_3))^3]'$. Clearly, the left inequality imposes $(u_1 - u_2, \mu)_{\Gamma_3} \leq (u_1 - u_2, \lambda)_{\Gamma_3}$ and so $(u_1 - u_2, \mu)_{\Gamma_3} = 0$ for each μ in $[(H_{00}^{1/2}(\Gamma_3))^3]'$, thus the continuity constraint is satisfied by the solution of the saddle-point problem.

The right inequality implies: $I_1(u_1) + I_2(u_2) \leq I_1(v_1) + I_2(v_2)$ for each (v_1, v_2) in $(H^1(\Omega))^3$, which means that u_1 and u_2 minimize the sum of the energies on Ω_1 and Ω_2 among the fields satisfying the continuity requirement, and so u_1 and u_2 are the restrictions to Ω_1 and Ω_2 of the solution of the primal problem (4).

The classical problem of the saddle-point problem (5) leads to the equations:

$$
\begin{cases}
A_1 u_1 + B_1^* \lambda = f_1 & \text{in } \Omega_1 \\
u_1 = 0 & \text{on } \Gamma_0 \cap \Gamma_1 \\
A_2 u_2 - B_2^* \lambda = f_2 & \text{in } \Omega_2 \\
u_2 = 0 & \text{on } \Gamma_0 \cap \Gamma_2 \\
B_1 u_1 - B_2 u_2 = 0 & \text{on } \Gamma_3
\end{cases}
\tag{6}
$$

A_1 and A_2 are the differential operators of the linear elasticity equations on Ω_1 and Ω_2 with Neumann boundary conditions on Γ_3, and B_1 and B_2 the trace operators over Γ_3 of functions belonging to V_1 and V_2.

The analysis of these equations shows that the Lagrange multiplier λ is in fact equal to the interaction forces between the substructures along their common boundary. Clearly, to get independent local displacement problems, it is necessary to introduce the forces on the interfaces. From a structural analysis point of view this is hardly a surprise. Nevertheless, the precise functional analysis above is useful because it allows classical results about finite element approximations for hybrid or mixed differential equations to be used.

4.2. Discretization of the Hybrid Formulation

A discretization with finite elements of the hybrid formulation (6) leads to the following set of linear equations in which the notations of variables associated to discrete problems are the same as the ones formerly used for the continuous formulation.

$$\left\{ \begin{array}{l} \mathbf{K}^{(1)}\mathbf{u}_1 + \mathbf{B}_1^t\lambda = \mathbf{f}_1 \\[2mm] \mathbf{K}^{(2)}\mathbf{u}_2 + \mathbf{B}_2^t\lambda = \mathbf{f}_2 \\[2mm] \mathbf{B}_1\mathbf{u}_1 - \mathbf{B}_2\mathbf{u}_2 = 0 \end{array} \right. \tag{7}$$

By elimination of the displacements in the equation (7), the problem can be written with respect to λ only:

$$\left[\mathbf{B}_1\ \mathbf{K}^{(1)^{-1}}\ \mathbf{B}_1^t + \mathbf{B}_2\ \mathbf{K}^{(2)^{-1}}\ \mathbf{B}_2^t \right] \lambda = \mathbf{B}_1\ \mathbf{K}^{(1)^{-1}}\ \mathbf{f}_1 + \mathbf{B}_2\ \mathbf{K}^{(2)^{-1}}\ \mathbf{f}_2$$

So λ satisfies the following equation:

$$\mathbf{D}\ \lambda = \mathbf{b} \tag{8}$$

with

$$\mathbf{D} = [\mathbf{B}_1\ -\ \mathbf{B}_2] \begin{bmatrix} \mathbf{K}^{(1)^{-1}} & 0 \\ 0 & \mathbf{K}^{(2)^{-1}} \end{bmatrix} \begin{bmatrix} \mathbf{B}_1^t \\ \mathbf{B}_2^t \end{bmatrix} \quad \text{and} \quad \mathbf{b} = [\mathbf{B}_1\ -\ \mathbf{B}_2] \begin{bmatrix} \mathbf{K}^{(1)^{-1}} & 0 \\ 0 & \mathbf{K}^{(2)^{-1}} \end{bmatrix} \begin{bmatrix} \mathbf{f}_1 \\ \mathbf{f}_2 \end{bmatrix}$$

Obviously, the $\mathbf{D}$ matrix is symmetric positive. It is definite if the interpolation spaces chosen for $\mathbf{u}_1$ and $\mathbf{u}_2$ and λ satisfy the discrete Ladyzenskaia-Babuska-Brezzi condition.

To be able to use the standard approximation results for the mixed or hybrid formulations, it is necessary to find such finite element spaces that the discrete Ladyzenskaia-Babuska-Brezzi condition is uniformly satisfied according to h, the mesh size parameter.

But generally, checking the uniform Ladyzenskaia-Babuska-Brezzi condition for the discrete problem may be tough (see [11] and [12]). The finite elements used for the Lagrange multiplier must be associated with polynomials of one degree less than the ones used for the primal unknowns, as the Lagrange multiplier is homogeneous to some partial derivatives of the solution of the primal problem. When using the hybrid formulation as described above to get a domain decomposition method, the Lagrange multiplier is introduced just to enforce the continuity condition. The values of the discrete multiplier do not need to be a good approximation of the continuous interaction forces between the substructures.

So, satisfying the uniform Ladyzenskaia-Babuska-Brezzi condition is not necessary in this case. The discrete form of the continuity constraint: $\mathbf{v}_1 = \mathbf{v}_2$ on Γ_3, can be written simply: $\mathbf{v}_1 = \mathbf{v}_2$ for each degree of the freedom located on Γ_3.

With such a condition, the discrete $\mathbf{B}_i$ matrices are just boolean restriction matrices. Taking this approximation is equivalent to have finite elements associated with the same polynomials for the Lagrange multiplier λ and for the displacements fields, and to take a collocation approximation for the integral:

$$\int_{\Gamma_3} (\mathbf{v}_1 - \mathbf{v}_2)\, \mu\, d\gamma .$$

The uniform Ladyzenskaia-Babuska-Brezzi condition for the discrete problem is not satisfied. But the displacement fields $\mathbf{u}_1$ and $\mathbf{u}_2$, solution of the discrete hybrid problem, are conforming, due to the condition: $\mathbf{u}_1 = \mathbf{u}_2$ for each degree of freedom located on Γ_3. So these fields are, in fact, the restriction over the two subdomains of the solution of the discrete primal global problem, for which the standard approximation results apply.

As a consequence of this form of discretization, one can see that the method can be applied even though the boundary of the interface Γ_3 is not embedded in Γ_0, the part of the boundary of Ω with homogeneous Dirichlet conditions.

4.3. Solution of the Discrete Hybrid Problem

The problem (8) can be solved through the conjugate gradient method because it is possible to compute the product of the $\mathbf{D}$ matrix by a vector, although the matrix itself is never computed.

Let μ be a vector, computing the product $\mathbf{v} = \mathbf{D}\,\mu$ involves the following three steps.

Step one: computation of the matrix-vector product,

$$\begin{bmatrix} \mathbf{v}_1 \\ \mathbf{v}_2 \end{bmatrix} = \begin{bmatrix} \mathbf{B}_1^t \\ \mathbf{B}_2^t \end{bmatrix} \begin{bmatrix} \mu \\ \mu \end{bmatrix}$$

that is just a reordering operation because the $\mathbf{B}i$ matrices are boolean.

Step two: computation of the product,

$$\begin{bmatrix} \mathbf{w}_1 \\ \mathbf{w}_2 \end{bmatrix} = \begin{bmatrix} [\mathbf{K}^{(1)}]^{-1} \\ & [\mathbf{K}^{(2)}]^{-1} \end{bmatrix} \begin{bmatrix} \mathbf{v}_1 \\ \mathbf{v}_2 \end{bmatrix}$$

that means computing the solution of two independent local sets of linear equations associated with the local linear elasticity problems with Neumann boundary conditions on the interface Γ_3.

Step three: computation of the variation on the interface of the displacement fields w_1 and w_2,

$$v = [B_1, -B_2] \begin{bmatrix} w_1 \\ w_2 \end{bmatrix} = B_1 \, w_1 - B_2 \, w_2 \, .$$

Obviously, the main step is the second one, and it can obviously be performed in parallel, whereas only the step three involves interprocessor data transfers. So, this method leads to a parallel algorithm with the same kind of granularity with the Schur complement method.

The B_i matrices obtained with the discrete hybrid method presented in the previous section are boolean matrices. So the contribution D(1) of the subdomain number 1 to the dual interface matrix is equal to:

$$D^{(1)} = B_1 [K^{(1)}]^{-1} B_1^t = \begin{bmatrix} 0 & I \end{bmatrix} \begin{bmatrix} K_{11} & K_{13} \\ K_1^t & K_{33}^{(1)} \end{bmatrix}^{-1} \begin{bmatrix} 0 \\ I \end{bmatrix}$$

Let us note:

$$\begin{bmatrix} C_1 \\ C_3^{(1)} \end{bmatrix} \begin{bmatrix} K_{11} & K_{13} \\ K_1^t & K_{33}^{(1)} \end{bmatrix}^{-1} \begin{bmatrix} 0 \\ I \end{bmatrix}$$

Then $D^{(1)}$ is equal to the matrix $C_3^{(1)}$.

From the previous relation, one can see that the C_1 and $C_3^{(1)}$ matrices satisfy the following equations:

$$K_{11} \, C_1 + K_{13} \, C_3^{(1)} = 0$$

$$K_{31} \, C_1 + K_{33}^{(1)} \, C_3^{(1)} = I$$

Hence, by elimination of the C_1 matrix in the equations above, we can derive the following equality:

$$. K_{31} \, K_{11}^{-1} \, K_{13} \, C_3^{(1)} + K_{33}^{(1)} \, C_3^{(1)} = \left(K_{33}^{(1)} - K_{31} \, K_{11}^{-1} \, K_{13} \, C_3^{(1)} \right) = I$$

So, the $D^{(1)}$ matrix is, in fact, the inverse of the Schur complement matrix $S^{(1)}$.

From a functional analysis viewpoint, the hybrid method is the dual method of the Schur complement method, because the condensed problem with the hybrid method is related to the forces on the interface, when the Schur complement operator is related to the displacement fields on the interface.

From the linear algebra viewpoint, with the discretization presented here, the duality of the two methods is simply represented by the following relation between the two interface operators:

$$\mathbf{D}^{(i)} = \left[\mathbf{S}^{(i)} \right]^{-1}$$

4.4. Topology of the Interface for Conforming and Non-Conforming Domain Decomposition Methods

There are some features of the interface topology which could make the domain decomposition method with Lagrange multiplier more suitable for a parallel implementation than the conforming Schur complement method.

When a point belongs to several subdomains, the coefficients of the condensed matrix for the degrees of freedom associated with this point will be the sum of the contributions of the various subdomains to which the point belongs. As regards data dependency within the context of the implementation of the method on multi-processor systems, each processor performing the computation associated with one subdomain, it means that the result of the product by the Schur complement matrix for such nodes will depend on more than two local contributions.

In a distributed memory context it means that subdomains are neighbouring from the moment that they have just one common node. In the case of a chessboard decomposition of a two-dimensional domain, each subdomain has eight neighbours. For real three-dimensional topology, the number of neighbours can be very large. For a decomposition in cubes, for instance, there would be as many neighbours as the sum of the number of faces, edges and vertices of a cube that is 26.

In a shared memory context, there is no problem with data transfers, but the assembly of the result of the product by the Schur complement matrix is still complex, due to the fact that the number of local contributions depends on the location of the point.

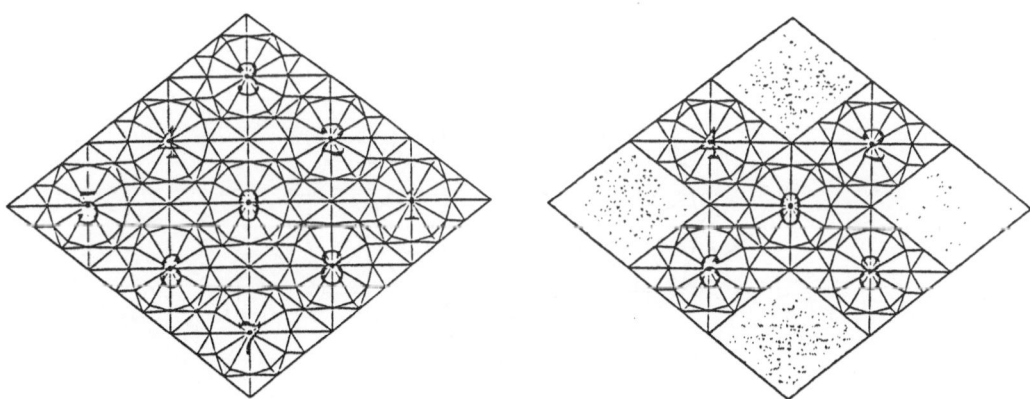

Figure 4. Neighboring domains with the Schur complement method and the hybrid method.

Let us now consider the domain decomposition method with Lagrange multiplier. The only sequential part of the computation of the product of the dual matrix $\mathbf{D}$ by a vector lies in the third step consisting in the computation on each interface of the variation of the displacement fields:

$$[\mathbf{B}_1 - \mathbf{B}_2] \begin{bmatrix} \mathbf{w}_1 \\ \mathbf{w}_2 \end{bmatrix} = \mathbf{B}_1 \, \mathbf{w}_1 - \mathbf{B}_2 \, \mathbf{w}_2 \;.$$

The $\mathbf{B}_1$ and $\mathbf{B}_2$ matrices are the discrete operators associated with the weak formulation of the continuity of the displacement fields on the interface:

$$(\mathbf{v}_1 - \mathbf{v}_2, \mu)_{\Gamma_3} = 0 \text{ for each } \mu \text{ in } [(H^{1/2}_{00}(\Gamma_3))^3]'$$

If the interface Γ_3 between two subdomains has a zero integral, this equation vanishes. In fact $\mathbf{B}_1$ and $\mathbf{B}_2$ are discrete trace operators, and the continuous trace operators are defined only on subsets of the boundary with non zero integrals. This is still true even though taking the same degrees of freedom for the Lagrange multiplier as for the restrictions over the interface of the displacement fields, as was presented in the previous section.

For each degree of freedom located on a point belonging to more than two subdomains, there are then as many degrees of freedom for the Lagrange multiplier as pairs of subdomains whose interface has a non-zero integral.

In the case of a chessboard composition of a two-dimensional domain, each subdomain has only four neighbours, one for each edge. Each degree of freedom for the displacement fields located at a vertex is associated with four degrees of freedom for the Lagrange multiplier, one for each edge intersecting at the vertex. For a decomposition in cubes of a three-dimensional domain, there are as many neighbours as faces, i.e. only six.

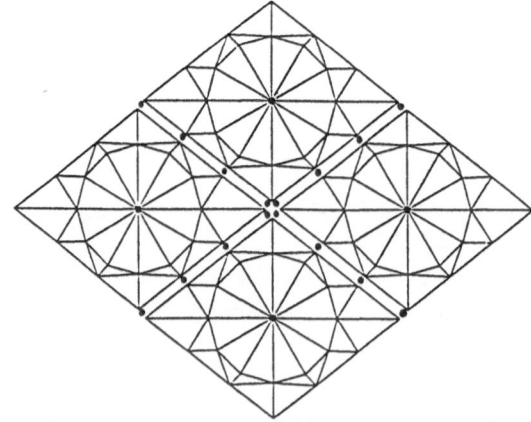

Figure 5. Degrees of freedom of the Lagrange multiplier for intersecting edges.

For implementation on a parallel system with local memory, it means that the number of processors connected to each node of the system needs to be equal to four for a two-dimensional splitting, with the same topology as for a finite difference grid, and equal to six for three-dimensional decomposition, still with the topology of a three-dimensional regular grid. In both cases the number of neighbours is obviously minimal.

In both shared memory or distributed memory contexts, the assembly of the product of a vector by the dual interface matrix **D** is simpler because all the points have the same status, and there are exactly two local contributions to the computation of the product for all the interface nodes.

5. IMPLEMENTATION OF THE HYBRID METHOD FOR SOLVING A THREE-DIMENSIONAL STRUCTURAL ANALYSIS PROBLEM

5.1. Presentation of the Problem

The test problem we consider consists in solving the linear elasticity equations for a composite beam made of little more than one hundred stiff carbon fibers bound by an uncompressible elastomer matrix.

Homogenization methods do not work for such a device with macroscopic-scale discontinuity and very different materials. For instance the Young modulus in the direction of the axis of the beam is 53000 MPa for the fibers and 7.8 MPa for the elastomer. But, due to the composite feature, the finite element mesh for solving the problem with discontinuous coefficients must be very refined, for it must discern the material discontinuity. That leads to a very large matrix, so the problem can be solved only through iterative methods like the conjugate gradient method.

However, substructuring is very easy in the present case for the beam is made of similar jointed composite "pencils" consisting in one of the fibers with its elastomer matrix.

Furthermore, it must be noticed that the problem we tackle is very ill conditioned. First, for geometrical reasons when trying to solve the pure bending problem, i.e. the case of a fixed bottom of the beam and transverse stresses on the top. The condition number

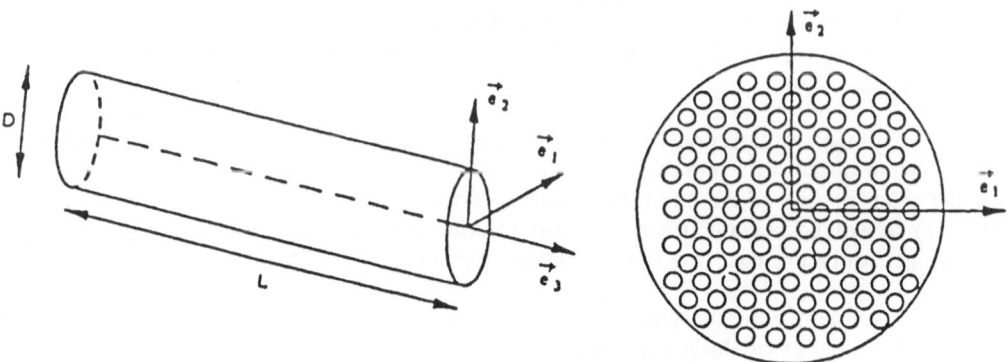

Figure 6. Geometry of the composite beam.

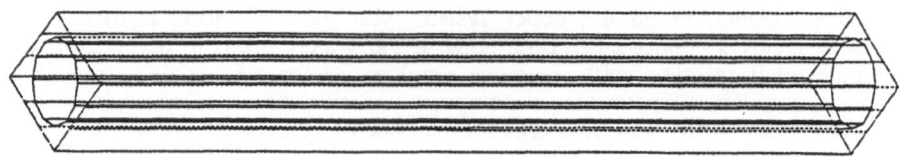

Figure 7. A composite "pencil".

of the matrix of the problem increases with the ratio of the length of the beam upon the width. This is exactly the case we are most interested in solving. Secondly, for material reasons, because of the composite feature, and because of the uncompressibility of the elastomer. To enforce this constraint, we introduce a penalty parameter and the condition number increases when this parameter tends to 0.

So, this problem gives a very good example of a stiff mechanical engineering problem with natural substructuring. For both the ill conditioning and the high dimension, only supercomputers allow to tackle it. For the same reason and because substructuring is straightforward, it is a very interesting problem for testing domain decomposition methods on parallel supercomputers.

Moreover, it is simple to build smaller test problems with the same features in solving the elasticity equations in domains made of only a few pencils, with the same global ratio of the length upon the width than for the complete beam.

5.2. Choice of the local solver

The first tests of the hybrid finite element method has been performed with the solution of the independent local subproblems through the conjugate gradient algorithm.

According to expectations, the hybrid method gave good results under the parallelism point of view. But comparison with the conforming global conjugate gradient method with a parallel matrix-vector product led to the conclusion that the non-conforming hybrid method was generally much more expensive, although the conforming conjugate gradient is less efficient for parallel processing.

The reason is clear: the ratio of the length upon the width is higher for one pencil than for a beam made of several pencils. Thus, substructuring leads to local problems with condition number greater than the one of the conforming primal problem. Then the choice of the conjugate gradient method for solving the local equations yields to a more expensive algorithm.

Clearly, the solution consists of using a direct method for solving the local problems. It is possible because the number of degrees of freedom in the substructures can be much smaller than the one of the complete domain. Furthermore, as each local set of equations needs to be solved several times, the time for the LU decomposition is not predominant as it is generally the case for direct solution methods.

This problem with the condition number of the local matrices seems to be linked with special features of the particular problem we try to solve. But the crucial point with domain decomposition methods is the convergence speed of the outer iterative process, because each iteration requires the solution of all local problems and, thus, is very expensive.

Thus, it is better to locate the interfaces in regions where the solutions are smooth. And as a consequence the ill conditioning due to geometry may well be worse in the subdomains.

Moreover, iterative methods are sometimes faster than direct methods because of the cost of the LU factorization of the matrices. But when there are many right hand sides, and it is the case with domain decomposition methods because of the outer iterative procedure, direct methods are generally more effective.

Furthermore, the domain splitting may be performed in such a way that the substructures have a slender shape, in order to get small bandwidths for the local matrices. So, the cost of the computation of the Choleski factorization of all the local matrices is much lower, in both CPU time and memory requirements, than for the matrix of the complete problem.

At last, it is clear that the use of iterative methods for solving the local problems prevents optimal load balancing because the number of operations cannot be forecast.

The hybrid method with solution of the local problems through the Choleski method has been implemented on CRAY-2 and CRAY-YMP832 machines for the three-dimensional problem presented in the previous section of the paper.

Tests have been performed with subdomains consisting of one or of a few pencils and numbers of subdomains between four and sixteen.

With highly optimized backward and forward substitutions for the local problems (see [13]), speed-ups have been more than 3 on CRAY-2 and more than 7 on CRAY-YMP. These performances are nearly the best possible, due to the problems with memory contention on these machines. The global computation speeds have been 700 Megaflops on CRAY-2 and a little less than 2 Gigaflops on CRAY-YMP.

These results prove the ability of algorithms based on domain decomposition methods with solution of the local problems through direct solvers to yield maximum performances with vector and parallel supercomputers.

5.3. Some comparisons of the performances of the hybrid domain decomposition method and the global Choleski factorization

Table 1 presents some results obtained with a direct global solver, the Choleski factorization, and the hybrid domain decomposition method, for three different test problems.

In all the cases, we computed the results of the pure bending problem for three-dimensional beams, consisting of four, nine, or sixteen composite pencils. We indicate the number of subdomains, each subdomain made of one pencil,
the global number of degrees of freedom, the number of degrees of freedom on the interface, and the CPU times and memory requirements. The Poisson ration for the elastomer is 0.49. The CPU times include the time for the Choleski factorization of all the local matrices for the hybrid domain decomposition method.

Table 1

number of subdomains	4	9	16
global number of d.o.f.	13000	28000	48000
hydrid domain decomposition method			
number of interface d.of.	2450	7350	14700
number of iterations	130	210	300
cpu time(s)	20	73	193
memory size (mw)	1.6	4.5	9.5
global Choleski factorization			
cpu time(s)	15	130	650
memory size (mw)	3.6	16	50

The stopping criterion for the outer conjugate gradient algorithm is:

$$|u_n - u| \, / \, |u| \leq 10^{-8}, \text{ and } |Ku_n - b| \, / \, |b| \leq 10^{-4} ,$$

where u is the result obtained with the global Choleski factorization, and K the global stiffness matrix.

These tests show that the domain decomposition method may be, with a well suited splitting of the domain, more efficient than the direct solution, from the points of view of both memory requirements and CPU time, even for very ill conditioned three-dimensional problems.

Furthermore, the domain decomposition method is much better suited for parallel processing, and it can be efficiently implemented on distributed memory machines, because the main part of the data lies in the LU decomposition of the matrices of the local problem that can be located in local memories. The data transfers involve only the traces of the fields on the interface, and so, are several orders of magnitude smaller than the number of operations to be performed in parallel for solving the local sub-problems (see [14] for a parallel implementation on an Intel hypercube machine of a preconditioner based on the Schur complement method).

6. CONCLUSIONS

The Schur complement hybrid domain composition methods appear, in practice, to have mixed characteristics of direct and iterative solution methods. They are iterative methods, because they consist of solving an interface problem through the preconditioned conjugate gradient method. Like other iterative methods, they entail lower memory filling than direct methods, because only the **LU** factorization of small local matrices are to be stored.

But with domain decomposition methods, the dimension of the problem to be solved through an iterative method is much smaller than the dimension of the complete problem. And the matrix of the condensed problem on the interface is much denser than the usually sparse matrix of the complete problem. Also its condition number is lower, because the elimination of the variables associated with the internal nodes represents some kind of block Jacobi preconditioner.

These characteristics make these domain decomposition methods, when using direct local solvers, much more robust than standard iterative methods. They represent a good way to use direct solvers for problems with such large numbers of degrees of freedom that the solution of the complete problems through a **LU** factorization would not be affordable.

The tests presented in the previous section show that domain decomposition methods can, with a well-suited splitting of the domain, be less expensive than direct solvers in both CPU time and memory requirements, even for very ill-conditioned three-dimensional structural analysis problems.

Furthermore, these methods lead to a very high degree of parallelism, and they are very well suited for being implemented on parallel systems with local memories, like distributed memory or hierarchical memory machines.

An open question that needs to be solved to make such algorithms general purpose solvers lies in the problem of mesh splitting and interface localization. Finding an optimal substructuring requires to take into consideration different problems. The subdomains must have such a shape that the local matrices have a low bandwidth in order to make the use of direct local solvers efficient. This may lead to large interfaces. But, the less points there are on the interface, the smaller the dimension of the dual problem; this should be better to ensure fast convergence of the outer conjugate gradient.

Furthermore the condition numbers of the local problems and of the dual problem depend upon the aspect ratio of the substructure and of the interface. To get round the local ill-conditioning, the use of direct local solvers and a reorthogonalization process for the outer conjugate gradient (see [15]) seem to be effective. But the repercussions for the condition number of the dual interface operator of the geometry of the decomposition are difficult to anticipate, because they depend not only on the aspect ratio of the interface but also on the mechanical features of the global problem.

REFERENCES

1. P.L. Lions, "On the Schwarz alternative principle", in *Proceedings of the First International Symposium on Domain Decomposition Methods for Partial Differential Equations, Paris, 1987*, R. Glowinski, G.H. Golub, G. Meurant and J. Periaux, eds., SIAM, Philadelphia (1988).
2. T.F. Chan and D. Goovaerts, "Schwarz = Schur: overlapping versus non-overlapping domain decomposition", to appear in *SIAM J. Sci. Stat. Comp.*
3. P.L. Lions, "On the Schwarz alternative principle, III: a variant for non-overlapping subdomains", to appear in *Proceedings of SIAM Fourth International Conference on Domain Decomposition Methods, March 20-22, 1989, Houston, Texas.*
4. P.E. Bjordstad and O.B. Wildlund, "Iterative methods for solving elliptic problems on regions partitioned into substructures", *SIAM J. Numer. Anal.*, Vol. 23, No. 6 (December 1986).

5. V.I. Agoshkov, "Poincaré's-Steklov operators and domain decomposition methods", in *Proceedings of the First International Symposium on Domain Decomposition Methods for Partial Differential Equations, Paris, 1987*, R. Glowinski, G.H. Golub, G. Meurant and J. Periaux, eds., SIAM, Philadelphia (1988).

6. P. Le Tallec, J.F. Bourgat, R. Glowinski and M. Vidrascu, "Variational formulation and conjugate gradient algorithm for trace operator in domain decomposition methods", in *Proceedings of the Second International Symposium on Domain Decomposition Methods for Partial Differential Equations, Los Angeles, 1988*, T.F. Chan et al., eds., SIAM, Philadelphia (1989).

7. Y.H. De Roeck, "A local preconditioner in a domain decomposed method", CERFACS Report TR89/10, March 1989, 31057 Toulouse Cedex, France.

8. T.H.H. Pian and P. Tong, "Basis of finite element methods for solid continua", in *Internat. J. Numer. Methods Engrg.*, Vol. 1, 3-28 (1969).

9. T.H.H. Pian, "Finite element formulation by variational principles with relaxed continuity requirements", in *The Mathematical Foundation of the Finite Element Method with Applications to Partial Differential Equations*, Part II, A.K. Aziz, ed., Academic Press, New York (1972), 671-687.

10. F. Brezzi, "On the existence, uniqueness and approximation of saddle-point problems arising from Lagrangian multipliers", *R.A.I.R.O., Analyse Numérique*, 8, R2 (1974).

11. P.A. Raviart and J.M. Thomas, "Primal hybrid finite element methods for 2nd order elliptic equations", *Math. of Comp.*, Vol. 31, number 138, 391-413.

12. V. Girault and P.A. Raviart, *Finite Element Methods for Navier-Stokes Equations*, Springer Verlag, Berlin (1986).

13. F.X. Roux, "Tests on parallel machines of a domain decomposition method for a structural analysis problem", in *Proceedings of the 1988 International Conference on Supercomputing, St. Malo, France, 1988*.

14. D. Goovaerts and R. Piessens, "Implementation of a domain decomposed preconditioner on an iPSC/2", *Parallel Computing 1989*, D.J. Evans and F.J. Peters, eds., Elsevier, Amsterdam (1989).

15. F.X. Roux, "Acceleration of the outer conjugate gradient by reorthogonalization for a domain decomposition method with Lagrange multiplier", to appear in *Proceedings of SIAM Fourth International Conference on Domain Decomposition Methods, March 20-22, 1989, Houston, Texas*.

TERPSICHORE: A THREE-DIMENSIONAL IDEAL MAGNETOHYDRODYNAMIC

STABILITY PROGRAM

David V. Anderson[1], W. Anthony Cooper[2], Ralf Gruber, Silvio Merazzi and
Ulrich Schwenn[3]

GASOV-EPFL, CH-1015 Lausanne, Switzerland

ABSTRACT

The 3D ideal magnetohydrodynamic (MHD) stability code TERPSICHORE has been
designed to take advantage of vector and microtasking capabilities of the latest generation
CRAY computers. To keep the number of operations small most efficient algorithms have
been applied in each computational step. The program investigates the stability properties of
fusion reactor relevant plasma configurations confined by magnetic fields. For a typical 3D
HELIAS configuration that has been considered we obtain an overall performance of 1.7
Gflops on an eight processor CRAY-YMP machine.

1. INTRODUCTION

On the way towards a thermonuclear fusion reactor there are several technological and
physical uncertainties to be understood and solved. One of the most fundamental problems is
the appearance in the machines of many sorts of instabilities which can either enhance the
energy outflow or even destroy the magnetic confinement of the fusion plasma. The
knowledge of such instabilities is a prerequisite to a good understanding of the behaviour of
actual experiments, and to the design of new devices. Most of the effort is devoted to the
study of axisymmetric toroidal configurations such as tokamaks or spheromaks and to
helically twisted toroidal devices such as stellarators.

One of the main objectives of the present fusion program is to show that present day
tokamaks such as TFTR (Tokamak Fusion Test Reactor, USA) or JET (Joint European
Torus, EURATOM) can reach temperatures, confinement times and β values (= ratio between
plasma pressure and the energy invested in the magnetic field) relevant to fusion reactors. The
quantity β measures the efficiency of an electric power reactor based upon the tokamak
principle. Different methods can be used to increase the value of β such as additional heating
(neutral gas injection or radiofrequency heating), gas injection, decrease of the aspect ratio
(minor to major radius), and increase of elongation or triangularity of the plasma cross

---

[1] NMFECC, Livermore, USA
[2] CRPP-EPFL, Lausanne, Switzerland
[3] MP-IPP, Garching, BRD

Scientific Computing on Supercomputers II
Edited by J. T. Devreese and P. E. Van Camp
Plenum Press, New York, 1990

section. Unfortunately, pushing the value of β too high leads to a destruction of the plasma column which is called disruption, or to a degradation of the confinement time. It is thought today that there exist limits of β which depend on the geometry, the magnetic circuits and the properties of the fusion plasma.

The most violent global instabilities arising in timescales of microseconds are those described by the linear, ideal, magnetohydrodynamics (MHD) equations. These are obtained when the linearized motion of a magnetically confined plasma around its equilibrium state is studied. In this model, we neglect non-ideal effects such as the influence of finite resistivity, viscosity or kinetic effects. It is well suited to model reactor relevant plasmas having very high temperatures of the order of 10^8 K. For example, resistivity only alters the ideal results in timescales of the order of milliseconds, which is three orders of magnitude slower than the timescale of ideal modes. Even though the model considered is simple and non-dissipative and the resulting eigenvalue problem symmetric, it is difficult to find general characteristics of the eigenvalue spectrum.

During more than 10 years now, stability computations have been made with different computer programs [1-4] to find the highest value of the plasma beta in two-dimensional configurations. An important outcome of these parameter studies was what one calls now the *Troyon limit* [5,6]. This scaling law tells us that the optimal beta scales with the total current and the inverse aspect ratio, and, for a given current, is independent of the geometry. A change in geometry only enables one to increase the total current. A very good agreement with experimental results could be found [6], which suspects that the beta limits found by two-dimensional numerical experimentation are pertinent ones. To overcome these limits, three-dimensional configurations, such as stellarators or 3D tokamaks, are proposed.

The advent of powerful vector and multiprocessor supercomputers with large memories like the Cray-YMP and the Cray-2 has made the investigation of the magnetohydrodynamic (MHD) stability properties of fully three-dimensional (3D) plasma configurations confined with magnetic fields feasible. The computer code TERPSICHORE has been developed for this purpose and has been explicitly designed to take advantage of the vector and microtasking capabilities of these types of computers.

To successfully tackle the linear stability problem in three dimensions using numerical methods, the computer program that is devised must be very carefully designed because compared with the 2D predecessors such as ERATO [1] or PEST [2] the memory and the CPU requirements can be correspondingly much larger. To become a useful tool, the 3D stability code must also be very fast in order to have the capability of exploring the larger parameter space associated with the extra degree of freedom that the third dimension introduces. In the construction of TERPSICHORE we have greatly benefited from the accumulated experience developed in the design and operation of the 2D stability packages ERATO, PEST, PEST2 [3] and the code of Degtyarev et al. [4]. Thus, a specially adapted coordinate system is chosen and double Fourier expansions in the poloidal and toroidal angles and a finite hybrid element approach [5] are performed to optimally describe the eigensolution. In order not to include unnecessary modes a special expert-system-like program is used to select those poloidal and toroidal Fourier terms which contribute to the instability. An adaptive mesh is used in radial direction. This leads to minimum size matrices in the generalized eigenvalue problem. As a consequence, it is no longer the eigenvalue solver but rather the construction of the matrix elements that becomes the most expensive computational step. Special care has been taken to compute the double Fourier flux tube integrals of equilibrium quantities to construct each matrix element in the most optimal way. These features clearly distinguish this code from its 2D predecessors which were developed in the mid seventies and were partially designed to benefit only from the vector and memory capabilities of Cray-1 computers that were available at that time. As a result, the TERPSICHORE code, although intended to be used as a 3D MHD stability package, performs the 2D stability problem much more efficiently and rapidly than either ERATO or PEST. In more complicated 3D low magnetic shear stellarator configurations we achieve an overall performance of 1.7 Gigaflop/s on an eight processor Cray-YMP machine. The parallelization needed to achieve this high rate was entirely done by autotasking.

2. THE PHYSICS PROBLEM

The 2D stability packages have been employed extensively in the last 10-15 years in the design of axisymmetric tokamak devices such as the Joint European Torus (JET) and the Tokamak Fusion Test Reaction (TFTR) that have already been built and have operated successfully. They are presently being used to design reactor-like devices such as the International Test Experimental Reactor (ITER) which is an international collaborative effort to consider the next step in magnetic fusion energy research. The purpose of the ideal MHD stability codes is to determine the possible boundaries of operation in the plasma current and in the pressure that can be confined. As a testament to the sucess of these codes was the extraction of the Troyon limit [6] from computer simulations which was subsequently verified experimentally. It should be noted that a substantial fraction of the computational effort in magnetic fusion energy research realized on Cray computers in the last 10 years has been devoted to the 2D stability problem.

There are strong motivating factors to extend the stability analysis to 3D configurations. First, the experimental conditions in tokamaks display in many cases highly nonsymmetric internal magnetic structures. Second, external helical windings are incorporated in many tokamak designs for disruption control (which is a class of instability described by resistive MHD in the nonlinear phase rather than by the linear ideal MHD model) that can significantly alter the symmetry properties of the device. In the near term, a modification of the JET machine is planned to experimentally test this concept. Third, though present devices like JET have 32 toroidal magnetic field coils to minimize the magnetic ripple effects and force the configurations to be as close to axisymmetric as possible, the accessibility to the plasma and cost considerations constrain future larger tokamak designs to be built with a reduced number of toroidal coils that will as a result spoil the symmetry and increase the relevance of 3D calculations. Finally, a fourth important motivating factor, is the recent interest in - and relative success of stellarator configurations that employ external coils rather than transformer-induced plasma currents to generate the confining magnetic fields, and thus offer the attractive potential of steady state operation. These types of devices, of course, are inherently 3D in character.

To explore the problem of the linear MHD stability of confined plasmas, we start from a plasma equilibrium solution. Finding an **equilibrium configuration** implies solving the nonlinear set of equations

$$\nabla p = \mathbf{j} \times \mathbf{B}$$
$$\mathbf{j} = \nabla \times \mathbf{B} \tag{1}$$
$$\nabla \cdot \mathbf{B} = 0 .$$

Whenever in what follows the quantities p, $\mathbf{j}$ or $\mathbf{B}$ appear, they denote, respectively, the equilibrium pressure, the equilibrium current density and the equilibrium magnetic field. This equilibrium solution (1) is perturbed. The linearized ideal MHD equations.[5] describing the **stability behaviour of the plasma** can be written in variational form :

$$\delta W_p + \delta W_v - \omega^2 \delta W_k = 0 \tag{2}$$

where δW_p represents the potential energy in the plasma, δW_v represents the magnetic energy in the vacuum region that surrounds the plasma, δW_k represents the kinetic energy and ω^2 corresponds to the eigenvalue of the system.

In 2D, it is straightforward to rigourously demonstrate that the MHD equilibrium consists of nested magnetic flux surfaces. A similar proof does not exist in 3D. Consequently, as a first step, we have imposed the condition of flux surface nestedness in the underlying 3D equilibrium state we wish to investigate. We thus exclude cases with magnetic islands and internal separatrices. The potential energy can then be described as

$$\delta W_p = \frac{1}{2} \int \int \int dx^3 \left[C^2 + \gamma p \,|\, \nabla \cdot \xi \,|^{\,2} - D \,|\, \xi \cdot \nabla s \,|^{\,2} \right] \tag{3}$$

where ξ represents the perturbed displacement vector, and γ is the adiabatic index,

$$C = \nabla \times (\xi \times B) + \frac{j \times \nabla s}{|\nabla s|^2} \, (\xi \cdot \nabla s) , \tag{4}$$

$$D = \frac{2 \, (j \times \nabla s) \cdot (B \cdot \nabla) \, \nabla s}{|\nabla s|^4} , \tag{5}$$

where the radial variable s labels the magnetic flux surfaces within the plasma domain $0 \le s \le 1$ from the magnetic axis to the plasma-vacuum interface. The vacuum energy is described as

$$\delta W_v = \frac{1}{2} \int \int \int d^3x \, (\nabla \times A)^2 \tag{6}$$

where A is the perturbed vector potential. There are two methods to treat this part of the problem. One is to employ a Green's function technique and the other is to consider the vacuum as a pseudoplasma and in such case the structure of the resulting matrix equation is virtually identical to that inside the plasma. The applications discussed in this paper are limited to internal plasma instabilities to which δW_v does not contribute, thus further details about the vacuum treatment are omitted.

A convenient form to represent the perturbed displacement vector is

$$\xi = \sqrt{g} \, \xi^S \nabla \theta \times \nabla \phi + \eta \frac{(B \times \nabla s)}{B^2} + \left[\frac{J(s)}{\Phi'(s)B^2} \eta - \mu \right] B \tag{7}$$

where θ and ϕ are the poloidal and toroidal angles, respectively, of the Boozer magnetic flux coordinate system which we have chosen for the stability analysis. This particular choice of coordinate system is optimal for the stability calculations because in its representation two very important conditions are satisfied. First, that the magnetic field lines are straight which allows a compact and efficient determination of the $B \cdot \nabla$ operator and thus an accurate description of the resonant surfaces. Second, it allows the most precise description of the parallel current (which is an important driving mechanism of instabilities in 3D systems) because the poloidal and toroidal magnetic fields in the covariant representation are current fluxes. The radial, binormal and parallel components of the perturbation vector are given by ξ^S, η and μ, respectively. The Jacobian of the transformation from cylindrical to Boozer coordinates is denoted by $\sqrt{g}$, $J(s)$ is the toroidal current flux function, $F(s)$ is the toroidal magnetic flux function and prime(') indicates the derivative of a flux surface quantity with respect to the radial variable s.

Noting that the perturbation component μ appears only in the term $\nabla \cdot \xi$ of δW_p, we impose the incompressibility constraint to eliminate it algebraically from the problem. This, however, implies that correct instability growth rates can no longer be computed, but the points of marginal stability are unaffected. Although we abandon the determination of growth rates, this procedure offers the advantage of reducing the size of the stability problem because only two instead of three components of the perturbation have to be calculated. This leads in addition to a reduction of coupling effects and the advantage of increased accuracy in the determination of marginal stability points because the continuous spectrum becomes displaced. This allows the application of interpolation rather than extrapolation techniques in the calculation. This approach was previously adopted in the development of the PEST2 code [3] which considerably speeded up and improved the accuracy of marginal stability computations compared with the 2D PEST stability package [2].

We apply a Fourier series decompostion of the perturbation components ξ^S and η given by

$$\xi^S(s,\theta,\phi) = \sum_l s^{-q_l} X_l(s)\sin(m_l\theta - n_l\phi + \Delta) \qquad (8)$$

and

$$\eta(s,\theta,\phi) = \sum_l Y_l(s)\cos(m_l\theta - n_l\phi + \Delta) \qquad (9)$$

where l is an index that labels the mode number pair (m_l, n_l), Δ is a phase factor, and the exponent $q_l = 0$ for $m_l \neq 1$ or $q_l = 1$ for $m_l = 1$. Then to satisfy regularity conditions at the origin, we have $X_l(0) = 0$. We employ a finite hybrid element radial discretisation because radial derivatives act only on X_l. As result, the energy principle described in Eq. (2) reduces in the weak form to an eigenvalue problem of the form

$$Ax = \lambda Bx \qquad (10)$$

where $x = (X_l, Y_l)$, the eigenvalue is $\lambda = \omega^2$, the matrix A is symmetric and block diagonal. As a result of imposing the incompressibility constraint, it is convenient to choose the matrix B that represents δW_k to be the unit matrix. The Fourier decomposition of the perturbation components ξ^S and η leads to the appearance in the matrix elements of A of what we referred to in the introduction as double Fourier flux tube integrals. Extensive algebraic manipulations were performed to reduce the number of these integrals to a total of seven as they can represent a significant fraction of the computational effort for realistic configurations. Typically, they are of the form

$$C_{lk}^{(2)}(s) = 2L_s/4\pi^2 \int \int d\theta \, d\phi \, \sqrt{g} \sin(m_l\theta - n_l\phi + \Delta) \sin(m_k\theta - n_k\phi + \Delta)$$

$$C_{lk}^{(3)}(s) = 2L_s/4\pi^2 \int \int d\theta \, d\phi \, (g_{ss}/\sqrt{g}) \sin(m_l\theta - n_l\phi + \Delta) \sin(m_k\theta - n_k\phi + \Delta) \qquad (11)$$

$$C_{lk}^{(4)}(s) = 2L_s/4\pi^2 \int \int d\theta \, d\phi \, (g_{s\theta}/\sqrt{g}) \sin(m_l\theta - n_l\phi + \Delta) \sin(m_k\theta - n_k\phi + \Delta)$$

where the limits of integration are given by $0 \le \theta \le 2\pi$ and $0 \le \phi \le 2\pi/L_s$ and L_s corresponds to the number of periods of the instability structure in one toroidal transit. For most cases of interest, $L_s = 1$. The quantities g_{ss} and $g_{s\theta}$ are lower metric elements.

In the equations (10), a solution **x** represents a stable displacement if $\omega^2 > 0$, a marginal situation if $\omega^2 = 0$ and an unstable displacement if $\omega^2 < 0$. Specifically, the ultimate goal is to find stable configurations for which the plasma pressure is as high as possible.

3. THE ORGANIZATION OF TERPSICHORE

The stability code TERPSICHORE consists of 6 basic modules, categorised as

(a) Interface to the MHD equilibrium

(b) Reconstruction of the MHD equilibrium

(c) Mapping the MHD equilibrium

(d) Construction of the stability matrix elements

(e) Eigenvalue solver

(f) Analysis and diagnostics of the results

The philosophy that underlies the construction of the routine that interfaces TERPSICHORE with an MHD equilibrium code is to rely on the minimum amount of information that is necessary and then subsequently reconstruct the equilibrium. This information consists basically of the geometry as well as the poloidal and toroidal magnetic fluxes. Thus, from the 3D MHD equilibrium code VMEC [9], which we have used exclusively so far as the source of 3D equilibria with nested flux surfaces, we obtain the Fourier amplitudes of the inverse coordinates R (the distance from the major axis) and Z (the distance from the midplane) to specify the geometry of the configuration. We also obtain from it $\Phi'(s)$ and $\psi'(s)$ which are the radial derivatives of the toroidal and poloidal magnetic flux functions, respectively. The computational effort in the interface is negligible. The reconstruction of the MHD equilibrium, on the other hand, constitutes a large fraction of the computer time that is spent. The Jacobian and metric elements in the original coordinates of the MHD equilibrium code are developed. The periodic poloidal angle renormalisation function [9] is computed in each flux tube by solving a linear elliptic partial differential equation that results from the condition $\mathbf{j} \cdot \nabla s = 0$ using Fourier techniques. The current densities are calculated and the accuracy of the reconstructed equilibrium is tested. The coordinate system that is optimal for the computation of 3D MHD equilibria using inverse Fourier methods consists of that in which the spectrum of modes required to obtain the equilibrium to a specified accuracy is minimized. It does not, in general, coincide with the Boozer coordinate system which, as we have discussed earlier, is optimal for the stability analysis of 3D systems. Thus, we have to perform a mapping from the coordinates of the equilibrium to the Boozer coordinates. From the expansion of the magnetic field in the covariant representation in the equilibrium coordinates, we obtain an auxiliary function that is required for the mapping procedure [10]. We then calculate the Fourier amplitudes of Boozer coordinate quantities (i.e. R, Z, B^2, etc.) directly from information in the real space equilibrium grid. Further details about the mapping procedure we have implemented can be found in Ref. 7.

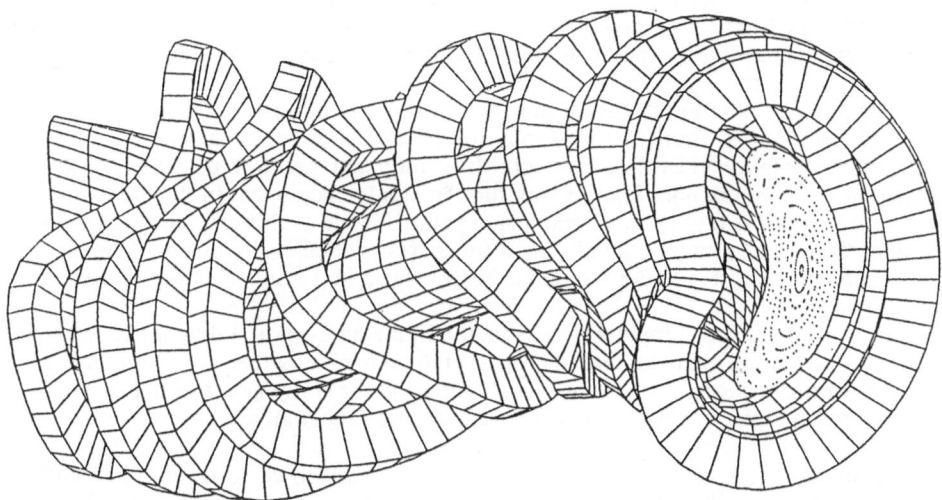

Figure 1. The magnetic field coil structure and the plasma shape in one field period of a 4 field period Helias stellarator configuration. The cross sectional cut shows the shape of the internal flux surfaces in the plasma obtained by puncture plots left by the intersection of magnetic field lines traced over many toroidal transits across the plane defined by the cut.

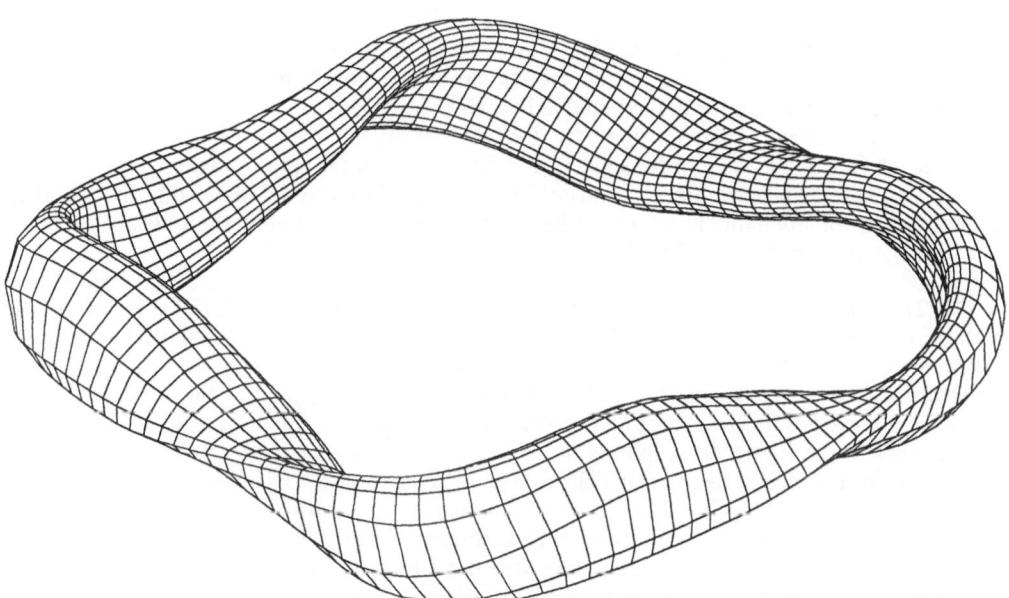

Figure 2. The full 3D structure of a flux surface corresponding to a 4 field period Helias stellarator configuration.

We then calculate the Jacobian and the metric elements that correspond to Boozer coordinates in real space, we test the accuracy of the equilibrium we have constructed in Boozer coordinates and develop additional quantities that are required for the stability analysis. This part usually constitutes a moderate fraction of the computational requisites.

We proceed next to the fourth step, which is the computation of the elements of the matrices A and B. We first read in a table of instability modes that are selected for the calculation. As mentioned earlier, the matrix B is unity and therefore trivial to construct. We separate out the seven double Fourier flux tube integrals that appear in the elements of A to design a routine that is highly vectorizable and parallelizable as this can represent a dominant fraction of the overall computational time if this part is not carefully treated. The eigenvalue solver inverts the matrix A to determine the eigenvalue and the eigenvectors using an inverse vector iteration method. This solver has been very carefully designed so that it uses only a very small portion of the computer time. Finally, in the analysis and diagnostic sections, the kinetic and potential energies are reconstructed, the eigenvalue is tested, the parallel component of the displacement vector is calculated and the components of the perturbation in real space are determined. The computational effort here is negligible.

In all these computational steps, we made a big effort in vectorization. In addition, the code is organized in such a way that in each module the computation for one radial interval be as independent as possible from the computations for the other radial intervals. This makes parallelization easy. Since the number of Fourier terms in each sequence of the execution is the same for each radial interval, the sequence of operations is the same as well. As a consequence, it will be possible in the future to run TERPSICHORE on a synchronous parallel SIMD computer, such as the Connection Machine.

4. THE TEST CASE

As a test case to investigate the performance and capabilities of the TERPSICHORE code, we have chosen a 4 field period Helias stellarator configuration [11] that has been considered as a candidate for the Wendelstein VII-X device that is being planned as the next step in stellarator experiments at the Max Planck Institut für Plasma Physik in the Federal Republic of Germany. As can be seen in the first two figures, the configuration we shall examine represents a particularly nontrivial test for the TERPSICHORE code. Its stability properties are not amenable to treatment by the stellarator expansion method because it has a helical magnetic axis, thus a fully 3D approach must be followed.

The coils that generate the toroidal and poloidal confining magnetic fields in one period of the device are shown in Figure 1. One can also perceive the shape of the plasma as it twists within the coil structure. The cross sectional cut at one end of the field period shown reveals the shape of the internal flux surfaces. To obtain these, several magnetic field lines at different radial locations were followed around the torus for a very large number of transits. Each point that appears in the figure corresponds to the intersection of one of the field lines with the vertical plane. Each field line that is followed thus yields a series of points on the plane that traces out a flux surface. The three dimensional character of a flux surface over the entire 4 period toroidal domain can be more clearly appreciated in Figure 2. The 3D MHD Helias equilibrium we investigate was generated with the VMEC code [9]. The volume averaged beta, which is a measure of the ratio between the plasma pressure and the confining magnetic field energy density was $\beta = 2\%$ for this case with vanishing net toroidal plasma current within each flux surface. The number of radial intervals was $N = 48$ and the number of modes needed to describe the equilibrium state was $N_e = 50$. In addition to the mode selection table, we obtain from VMEC the Fourier amplitudes R and Z, as well as $\Phi'(s)$ and $\psi'(s)$ on each flux surface. The number of modes required to reconstruct the equilibrium state in the Boozer coordinates was $N_b = 160$. Note that the Boozer spectrum is much broader than that of the

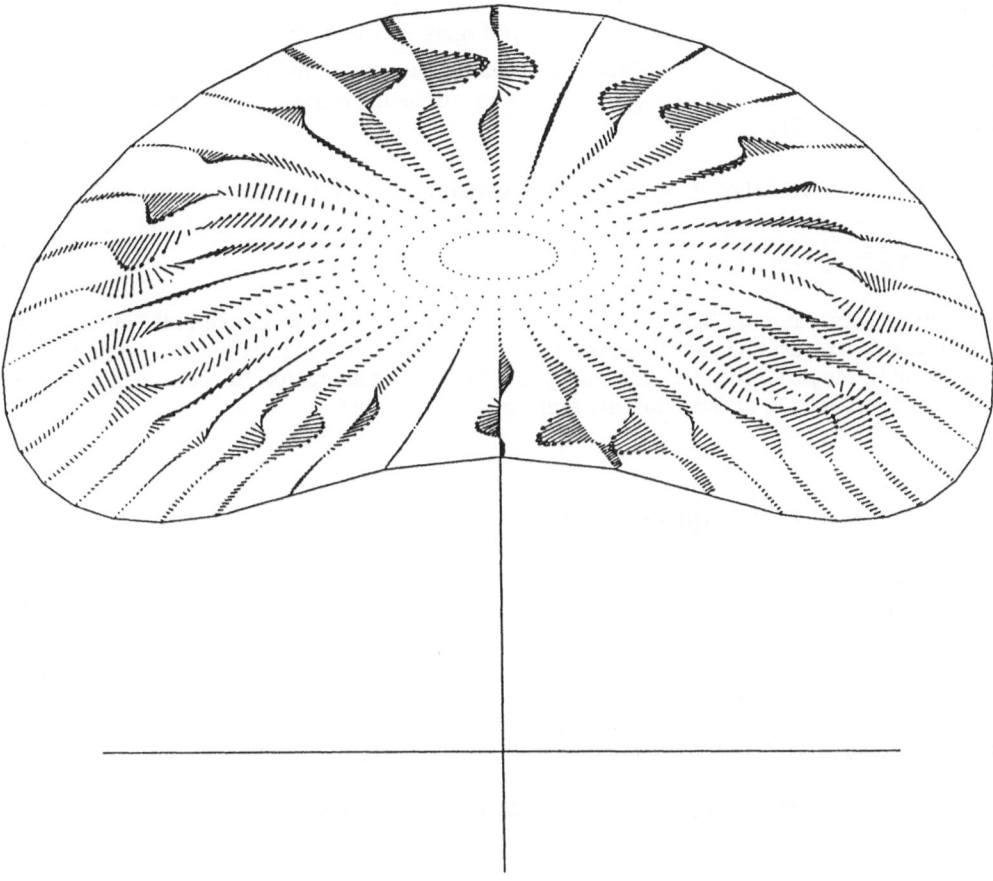

Figure 3. The flow pattern corresponding to the most unstable eigenmode structure at the toroidal plane with Boozer coordinate toroidal angle $\phi = 0$ in a 4 field period Helias stellarator configuration. The structure is dominated by an $m = 4$, $n = 3$ mode concentrated about the flux surface having a rotational transform of 0.75. The configuration has a plasma $\beta = 2\%$ and zero net toroidal plasma current within each flux surface.

original equilibrium coordinates. This demonstrates that the direct computation of the MHD equilibrium in the Boozer coordinates would be inefficient and possibly unfeasible. To prepare the computation of the Fourier integrals, the number of intervals in the poloidal angle variable and in the toroidal angle variable were chosen as $N_p = 72$ and $N_t = 32$, respectively.

The selection table of modes to describe the internal plasma instability structures for this device was chosen to have $N_s = 32$ modes. The stability analysis computed with TERPSICHORE demonstrated that this configuration was weakly unstable. The magnitude of the most unstable eigenvalue was $\lambda = -1.67 \times 10^{-3}$. The corresponding eigenmode structure on the toroidal plane that has $\phi = 0$ is shown in Figure 3. The dominant structure corresponds to an $m = 4$, $n = 3$ mode localized about the surface with rotational transform $\iota(s) \equiv d\psi/d\Phi = 0.75$. It should be noted that coupling effects to sidebands represent an important aspect in driving this instability.

On the toroidal plane with $\phi=0$, we show in Figure 4 the perturbed pressure

$$\delta p = -p'(s)\, \xi^s \tag{12}$$

in form of a colour plot. The perturbed pressure appears as convective cells in which the colour orange corresponds to the maximum value and the colour violet to the minimum value. On this plot, the instability shows a noticeable ballooning character. In addition, such colour plots for 128 different toroidal cuts are represented in Figure 5.

5. PERFORMANCE MEASUREMENTS

5.1. Operation counts

The operation counts for all important time consuming computational steps mentioned in chapter 3 give:

(b) *Reconstruction of the MHD equilibrium:*

$$O_r = 34\, N\, N_p\, N_t\, N_e + N\, N_p\, N_t\, N_b\, (6.5\, N_b + 75) + 2\, N\, N_b^3$$

(c) *Mapping of the MHD equilibrium:*

$$O_m = 144\, N\, N_p\, N_t\, N_b$$

(d) *Construction of the stability matrix elements:*

$$O_f = 14.5\, N\, N_p\, N_t\, P\, N_s^2$$

(e) *Eigenvalue solver :*

$$O_e = 20\, N\, N_s^3 + 38\, N\, N_s^2\, N_{it}\,,$$

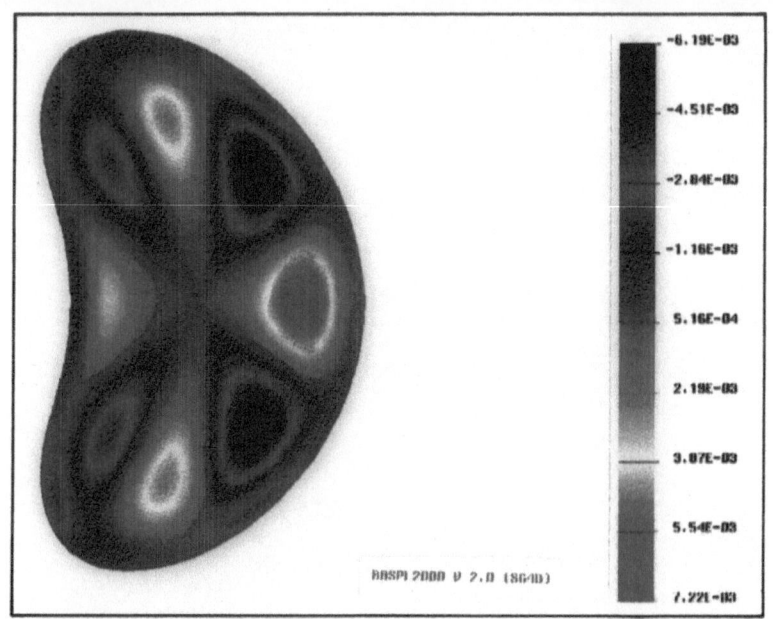

Figure 4. Colour plot of the perturbed pressure at the toroidal angle $\phi = 0$. The instability has ballooning character.

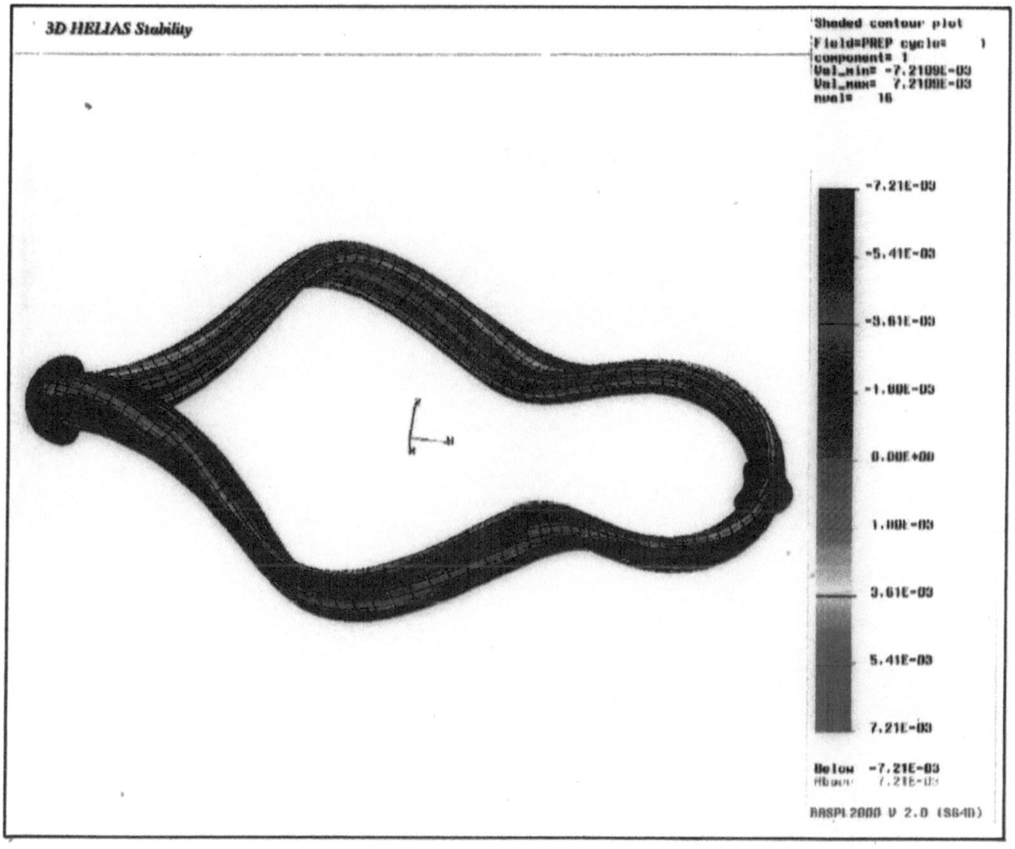

Figure 5. Perturbed pressure plots at different toroidal angles.

where

    N   Number of radial flux tubes

    N_p Number of poloidal points for the Fourier integrals

    N_t Number of toroidal points for the Fourier integrals

    N_e Number of Fourier terms in equilibrium

    N_s Number of Fourier terms in stability

    N_b Number of Fourier terms to represent equilibrium and geometrical quantities in Boozer coordinates

    P   Number of equilibrium periods

    N_{it} Number of inverse iteration steps in the eigenvalue solver

For the practical Helias case (N=48, N_p=72, N_t=32, N_e=50, N_s=32, N_b=160, P=4, N_{it}=11), the following operation counts are obtained:

$$O_r = 20.0 \times 10^9$$

$$O_m = 2.6 \times 10^9$$

$$O_f = 6.8 \times 10^9$$

$$O_e = 0.1 \times 10^9$$

leading to a total of 29.5×10^9 operations. The reconstruction of the equilibrium solution which includes the preparation for the most efficient evaluation of the Fourier integrals takes 68% of them, the mapping 9%,and the Fourier integrals 23%. The timing of the eigenvalue solver is negligible.

5.2. Parallelization procedure

Before attempting procedures that effect the parallelization of the TERPSICHORE code, the double Fourier flux tube integrals that appear in the subroutines in which the MHD equilibrium is reconstructed and in the subroutines in which the elements of the stability matrix A are determined are treated with successive applications of the Cray Research, Inc. mathematical routine MXM that performs matrix multiplications. Although this entails an increase in the number of floating point operations and in the storage requirements, the very efficient vectorization properties of the MXM routine reduces the computational time substantially. The bulk of the computational effort is concentrated in three subroutines. These calculate the periodic poloidal angle renormalization function, the mapping of the MHD equilibrium to Boozer coordinates and the elements of the stability matrix A. The time consumption in these subroutines is caused by repeated calls to MXM and to routines that evaluate trigonometric functions. We tried at first to parallelize only these three particular routines, and although notable improvements in performance were achieved, we deemed these to be unsatisfactory.

The optimal performance realized with the TERPSICHORE code was obtained with the application of Cray Research autotasking utilities through the activation of options available on the *cf77 -Zp* compiler to all the subroutines in the program with the exception of the interface to the MHD equilibrium data. For reasons as yet unknown, the attempt to autotask this routine that reads in the data caused the code to yield an incorrect eigenvalue. The computational time spent in this routine, however, is negligible. Consequently, a successful parallelization of the interface routine is not expected to improve the performance of the code in a significant way.

The design of TERPSICHORE always envisaged an eventual attempt to parallelize the program. It was originally our intention to try to multitask the most time consuming subroutines. However, the high level of parallelism achieved with the autotasking features available on the latest generation Cray computers reduces considerably the attractiveness of a multitasking approach because any further performance improvements that could be realized would be of very limited significance if not, in fact, counterproductive.

5.3. Timings

(a) Cray-2

The Helias case we have described above ($N=48$, $N_p=72$, $N_t=32$, $N_e=50$, $N_s=32$, $N_b=160$, $P=4$, $N_{it}=11$) was run on a 2 processor Cray-2 in unitasking mode. The original version of TERPSICHORE written purely in Fortran took 228 seconds which corresponds to a rate of 130 Mflop/s on one processor. The improved version on a dynamic memory Cray-2 takes 143 seconds in unitasking mode. This corresponds to 247 Mflop/s. It has also been run on a static memory four processor Cray2 using autotasking. A rate of 1.08 Gigaflop/s has been measured.

(b) Eight processor Cray-YMP parallelized

Performance tests of the TERPSICHORE code were also carried out on the sn1001 CrayYMP/832 computer. This machine has 8 processors, a 32 Megaword memory and a 6.41 ns clock period. The TERPSICHORE version used in the previous subsection was run first to produce the benchmark results. Subsequently, all calculations were performed with the improved version of the code in which all double Fourier flux tube integrals are treated with successive applications of the MXM routine. For the Helias case under consideration, the number of floating point operations increased to 3.49×10^{10}. The machine was dedicated to these tests. The results and performance of the code are summarized in the table below:

Table

	Original TERPSICHORE Unitasking	Improved TERPSICHORE Unitasking	Microtasking
CPU time	157.02"	137.06"	149.85"
Wall clock time		137.21"	20.44"
Parallelization	1	1	7.35
Gflop/s	0.189	0.255	1.708
Operations	29.6×10^9	34.9×10^9	34.9×10^9

Of the figures shown in the table, we would like to draw particular attention to the high level of parallelization of 7.35 in an eight processor machine achieved with the Cray Research autotasking utilities and to the 1.708 Gflop/s performance obtained.

Noting that the number of operations in the improved version of the code was increased relative to the original version, we can define an effective performance of 1.708x29.6/34.9=1.449 Gflop/s. On a YMP/8 machine with a cycle time of 6 ns instead of 6.41 ns, one expects 1.825 Gflop/s instead of 1.708 Gflop/s.

ACKNOWLEDGMENTS

We would like to express our gratitude to Stephen Behling of Cray Research Inc., to Arno Liegmann and to Jean-Pierre Therre of Cray Switzerland for the time and efforts devoted to the successful parallelization of the TERPSICHORE code. We also wish to thank Dr. Carlos Marino of Cray Research and Michèle Neyret of Cray Switzerland for their encouragements.

REFERENCES

1. R. Gruber, F. Troyon, D. Berger, L.C. Bernard, S. Rousset, R. Schreiber, W. Kerner, W. Schneider and K.V. Roberts, ' ERATO stability code', *Comput. Phys. Commun.* 21:323-371 (1981).
   R. Gruber, S. Semenzato, F. Troyon, T. Tsunematsu, W. Kerner, P. Merkel and W.Schneider, 'HERA and other extentions of ERATO', *Comput. Phys. Commun.* 24:363-376 (1981).
2. R.C. Grimm, J.M. Greene and J.L. Johnson, ' Computation of the MHD spectrum in axisymmetric toroidal confinement systems', *Meth. Comput. Phys.* 16:253-280 (1976).
3. J. Manickam, R.C. Grimm, R.L. Dewar, ' The linear stability analysis of MHD models in axisymmetric toroidal geometry', *Comput. Phys. Commun.* 24:355-361 (1981).
4. L.M. Degtyarev, S.Yu. Medvedev, ' Methods for numerical simulation of ideal MHD stability of axisymmetric plasmas ', *Comput. Phys. Commun.* 43:29-56 (1986).
5. R. Gruber, J. Rappaz, *Finite element methods in linear ideal MHD*, Springer Series in Computational Physics, Springer Verlag,Heidelberg (1985) ISBN 3-540-13398-4
6. F. Troyon, R. Gruber, H. Saurenmann, S. Semenzato and S. Succi, ' MHD limits to plasma confinement', *Plasma Phys.* 26:209-215 (1984).
   F. Troyon and R. Gruber, ' A scaling law for the beta-limit in Tokamaks', *Phys. Letters* 110A:29-34 (1985).
7. D.V. Anderson, W.A. Cooper, U. Schwenn and R. Gruber, ' Linear MHD stability analysis of toroidal 3-D equilibria with TERPSICHORE', in *Proceedings of the Joint Varenna-Lausanne International Workshop on theory of fusion plasmas*, Editrice Compositori, Bologna (1988), 93-102.
8. D.V. Anderson, A.R. Fry, R. Gruber and A. Roy, ' Gigaflop speed algorithm for the direct solution of large block-tridiagonal systems in 3-D physics applications', *Computers in Physics* 3:33-41 (1989).
9. S.P. Hirshman, W.I. van Rij and P. Merkel, ' Three-dimensional free boundary calculations using a spectral Green's function method', *Comput. Phys. Commun.* 43:143 (1986).
   S.P. Hirshman, U. Schwenn and J. Nührenberg, ' Improved radial differencing for 3D MHD equilibrium calculations', to be published in *J. Comput. Phys.* (1989).

10. J. Nührenberg, R. Zille, ' Equilibrium and stability of low-shear stellarators', in *Proceedings of the Workshop on theory of fusion plasmas,* Editrice Compositori, Bologna (1987), EUR11336EN, 3-23

11. J. Nührenberg, R. Zille, ' Stable stellarators with medium β and aspect ratio', *Phys. Letters* 114A:129-132 (1986).

THE BRIDGE FROM PRESENT (SEQUENTIAL) SYSTEMS TO FUTURE (PARALLEL) SYSTEMS: THE PARALLEL PROGRAMMING ENVIRONMENTS EXPRESS AND CSTOOLS

Patrick Van Renterghem

Automatic Control Laboratory, State University of Ghent (RUG), Grotesteenweg Noord 2, B-9710 Gent-Zwijnaarde, Belgium

ABSTRACT

Too many people still think that programming, debugging and porting programs to parallel programs is a very tough and almost impossible job. However, with the right tools, programming parallel systems can be almost as simple as programming sequential systems. One of the best environments for programming parallel computers is Express, based on ideas and developments from Caltech, one of the birthplaces of distributed memory machines. Basically, Express is a library of around 100 functions, which perform all the basic tasks of parallel computing: interprocessor communication, accessing hosts, sharing the parallel system, parallel graphics, configuration management, multitasking, performance analysis and debugging. The use of Express makes it possible to port sequential code to a parallel computer and it is extremely easy to port the resulting code to another parallel computer.

As an example, we discuss the implementation of a very computation-intensive neural network simulation on an arbitrary network of transputers. The solution presented can easily be applied to certain problems in theoretical and applied science such as computational fluid dynamics, stress analysis, ... By simply changing a command line argument, the program runs on 1, 2, 4 or even tens or hundreds of transputers.

We also compare this programming environment with another very popular parallel programming environment, CSTools from Meiko. Transputers have given a boost to parallel computing. Will these environments, or a combination of both, do the same for the development of software for these parallel computers?

INTRODUCTION

As the title suggests, this is a paper on future (parallel) computer systems and their programming environment. Even today, many computer systems have several processors, interconnected via local area networks or via special high-speed communication links. Within a few years, parallel processing will cease to be viewed as a weirdness and become a mainstream activity. The Inmos transputer, a revolutionary processor with a good performance as a single processor, but specially developed for building multiprocessor systems, has given a boost to parallel computing. The transputer has made it so easy to build a multiprocessor system with an enormous high potential performance that tens of companies are offering such systems today. Of course, to run an application on such a machine, the programmer must decompose his problem into smaller subproblems, which can be run on different processors to decrease the execution time of the application. This is where the problem lies. Traditional programming languages that were used to write an enormous amount of sequential code cannot deal with the problems posed by parallel computers, such as problem decomposition, data communication and synchronization. New languages are invented which are perfect for parallel processing, but we cannot expect everyone to rewrite his existing programs in this new and unfamiliar language.

Distributed memory machines have the highest potential for performance, but the hardest programming interface. Players in this rapidly growing area include the 80386-based Intel iPSC-2 hypercube and the transputer-based systems, made by more than 40 different companies. The transputer was specially designed for this kind of architecture and is obviously the most cost-effective building block for such systems. These systems fill the gap between superminicomputers (e.g. VAX 8600) and the vector-oriented super-computers, and bring parallelism to the people who cannot afford it, or can wait some-what longer for an answer. Systems are being built with 1024 transputer nodes (e.g. the Edinburgh Concurrent Supercomputer), which have a potential performance of several thousands of MIPS and several GFLOPS. Of course, traditional vector supercomputers have no problem to outperform a transputer system on problems that vectorize well, but the transputer system will be a lot (roughly 10 to 20 times) cheaper on initial cost and maintenance. A large transputer system costs approximately 50 kBEF per MFlop, whilst a large supercomputer costs something like 1 MBEF per peak MFlop. It is obvious that we will find these cost-effective machines more and more on the researcher's desktop. The combination of distributed processing and vector processing, as on Meiko's MK086 board or Inmos's B420 vector transputer module, enables us to build systems with a very good scalar and a very good vector performance. This is indeed a very interesting development.

The lack of convenient parallel programming environments is one of the major obstacles to the widespread use of multitransputer systems for general applications. It is great to have a cost-effective 100 MFlops machine, but what is it worth if it is very hard and time-consuming to program and debug your application on it? However, with the right tools, programming parallel systems can be almost as simple as programming sequential systems. What we need is a virtual machine interface (VMI) standard, which should be the same for sequential and parallel computers, for distributed and shared memory machines and for transputer-based and e.g. Intel-based machines. The VMI hides the hardware characteristics from the programmer, who can develop software for all machines with this standard programming interface. Of course, it would be possible to build a new, implicitly parallel programming language on top of this interface. The definition of such a VMI standard would be a very important step forward for the acceptance of parallel computers. We know that the companies behind two of the most popular environments of the moment, Parasoft's Express and Meiko's CSTools, have initiated talks to come to such a standard. In this paper, we briefly discuss both programming environments. The existence of such a standard will make parallel programming simple, efficient, portable and accessible to everyone.

REQUIREMENTS OF A GOOD PARALLEL PROGRAMMING ENVIRONMENT

The exploitation of the potential performance of the transputer system requires the efficient use of the hardware, load balancing and the optimization of data communication. Thus, a parallel programming environment should be efficient. If programmers can speed up an application considerably by bypassing the parallel programming environment, they will certainly do that and they will end up with programs, which are not portable, hard to read and hard to maintain. If such improvements exist, it is the task of the environment or compiler company to enhance the product.

People who wish to port large, existing programs to new parallel hardware would like to get an acceptable performance without having to rewrite their application. So, a parallel programming environment should be easy to use and the porting process should be cheap.

Furthermore, parallel programs should be independent of the interconnection and the number of nodes in the transputer system. A parallel programming environment should protect the user from needing intimate knowledge of the underlying hardware. It should be simple, transparent and reliable. Scalability and semi-automatic decomposition of data and processes that make programs independent of the number of processors in the system are equally important.

Furthermore, an enormous range of parallel systems is already available and many more will follow. We cannot be expected to rewrite our applications whenever we move from one system to another. Portability will even be a more important issue in the future when all machines are parallel machines.

Communication efficiency is extremely important in distributed memory machines, so the implementation of intercommunication and synchronization functions should be as efficient as possible. At the hardware level, a transputer can exchange medium-length messages at roughly 1.5 MB/s with an adjacent transputer, but you still need a through-routing mechanism to pass messages to non-adjacent processors (this will be done in hardware on the next-generation transputer, with code name H1 and a performance in excess of 100 MIPS and 20 MFlops, faster links and more on-chip memory). Of course, a general message-passing system always looses some of its efficiency, but the question is how much. Experiments have shown that this can be as high as 40 % in certain environments, which means that the compiler and software manufacturer can still improve their products substantially. We believe that this is a better attitude than trying to bypass the standard communication mechanisms and writing non-portable programs.

A good parallel programming environment should provide distributed I/O capabilities, if possible by implementing Unix-like I/O functionality.

Last but not least, a good parallel toolkit should also include a network debugger, to debug an application under development, and a performance monitor, to detect the bottlenecks in the implementation.

Personally, I think the three main requirements of a parallel programming environment are, in order of importance:

- ease of use
- efficiency
- portability

AN OVERVIEW OF PARALLEL PROGRAMMING ENVIRONMENTS AND LANGUAGES

It should be said that certain people are already very happy to be able to run their sequential program unchanged on a single transputer system, noticing that it runs 20 times faster than on an IBM PC/AT. Other people take advantage of the fact that the same algorithm can be applied to different sets of data and they are able to run this application on a multiprocessor machine without any effort. However, the majority of users needs to parallelize their application to obtain a drastic performance improvement.

Scepticists can no longer say that there are no programming environments for parallel computers. In fact, you could even say that too many have been developed, which makes it difficult to choose and difficult to port programs from one parallel system to another. There are a number of relatively new software approaches to parallel programming, which can be subdivided into programming languages, environments and operating systems.

New Programming Languages

- Linda from Yale University, marketed by Scientific Computing Associates - Parallel C compilers from 3L, Parsec and Definicon
- Parallel Fortran and Pascal from 3L Ltd.
- Strand88 from AI Ltd. and Strand Software Technology Inc.
- Transputer Ada from Alsys Ltd.

New Environments

- CSTools from Meiko Ltd.
- Express from Caltech, marketed by Parasoft
- SEP (System Environment Package) from Southampton University and Parseq - Trollius from Cornell University and PRE from Niche Technology Ltd

New Operating Systems

- Helios from Perihelion Software Ltd.
- Mach from Carnegie-Mellon University
- transIdris from Real Time Systems Ltd.
- V from Stanford University

The languages presented are either versions of conventional languages with extensions for parallel processing or new, implicitly parallel languages with interfaces to conventional programming languages.

Serious benefits are to be expected from automatic parallelizing compilers, although this can take a few years. An automatic parallelizing compiler analyses a sequential program and identifies the parallel sections (sections of code in which multiple threads of control can be active at the same time) and automatically generates parallel code for them. Only a limited class of parallel constructs can be found automatically:

- loop transformation: detection of DO loops that can be executed concurrently
- program segmentation: depending on a program flow analysis, we can identify independent segments of code which can be executed concurrently
- program reordering: reordering of program statements or loop nesting can make programs more viable for detecting parallel constructs.

Fully automatic parallelising compilers are still a long way off, so we will have to give the programmer all means of expressing and exploiting the parallel computer easily and efficiently. Furthermore, the best parallel algorithm for solving a particular application is often totally different from the best sequential algorithm. It should be emphasized that adjusted computer education and parallel algorithm teaching are vitally important here.

In the following paragraphs, we discuss 2 parallel programming environments, which have the intention to make parallel programming easier, more efficient and more portable. They work under familiar operating systems (Unix, DOS, ...), which makes the step from your familiar programming environment to the new parallel system smaller and easier.

THE CSTOOLS CROSS-DEVELOPMENT TOOLSET

CSTools is a program development toolset for multiprocessor systems. It supports conventional programming languages (C and Fortran) via compilers and configuration systems and provides run-time facilities such as high-level communication and symbolic debuggers. Using CSTools, application processes can be written entirely in C or Fortran, without having to learn a new language or new constructs.

CSTools runs on a Sun or a VAX, but also under MeikOS, Meiko's operating system for their Computing Surface multitransputer system, which was introduced in 1986. CSTools is not a new operating system for parallel machines, but a set of utilities for parallel programming. It is based on CSP (Communicating Sequential Processes), with sequential processes working together to perform a single overall task. To make use of

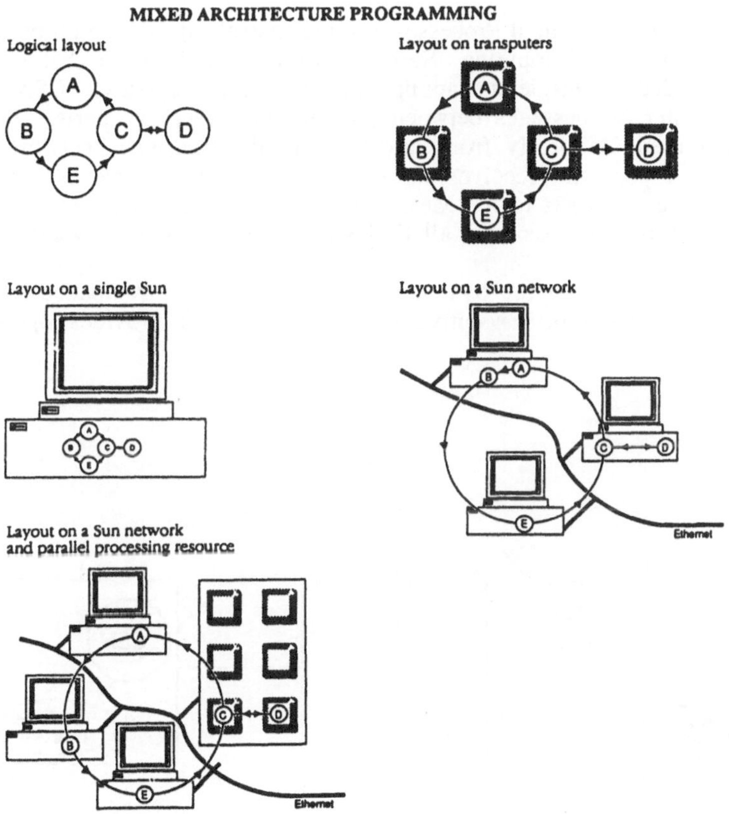

MIXED ARCHITECTURE PROGRAMMING

Logical layout

Layout on transputers

Layout on a single Sun

Layout on a Sun network

Layout on a Sun network
and parallel processing resource

Figure 1

the communicating sequential processes approach, the programmer must first design the application as a number of separate processes. Data exchange and synchronization is done via message passing routines. Running these different processes on different processors can dramatically decrease the execution time. CSTools does not really care if one of those sequential processes is running on a transputer or a Sun workstation. Neither does it care if communication happens via the links of the transputer, via the VMEbus of a Sun workstation or via a local area network. This makes it possible to develop applications which run on a mixed architecture system or on a single workstation (see Figure 1) without changing the source code.

The tdb symbolic debugger and the window-based tdbtool are derived from the standard Unix debugger dbx and Sun's dbxtool. When debugging multiprocessor and multitasking applications, one tdb session is invoked per process of interest.

The CSTools system has four primary components:
- a set of compilers for C and Fortran (mcc, mf77)
- a library of communication routines (system-global message ports, channels) - a loader for sequential and parallel programs (mrun)
- a runtime support environment (RTE, CSN)

Creating a parallel application using CSTools involves the following steps:
- the application is designed as a set of communicating sequential processes
- the individual sequential programs are written in C or Fortran and message passing library routines are used for communication and synchronization
- a PAR file, describing the distribution of component programs across processors (mapping) is written, loaded and run.

Communicating Sequential Processes wanting to exchange data with other processors, can do this via the Computing Surface Network (CSN) mechanism, which is automatically loaded alongside the application whenever necessary. CSN transparently handles the transfer of messages between system-global message ports. A port can be refered to and used identically from anywhere in the transputer network. Only one process creates a port and receives messages through it. However, any number of processes can send messages to it. Messages are automatically buffered and queued by the CSN. CSN forms the basis for all CSTools communications services.

Another Runtime Support System which is automatically loaded alongside user applications is the Runtime Executive (RTE). The RTE provides operating system

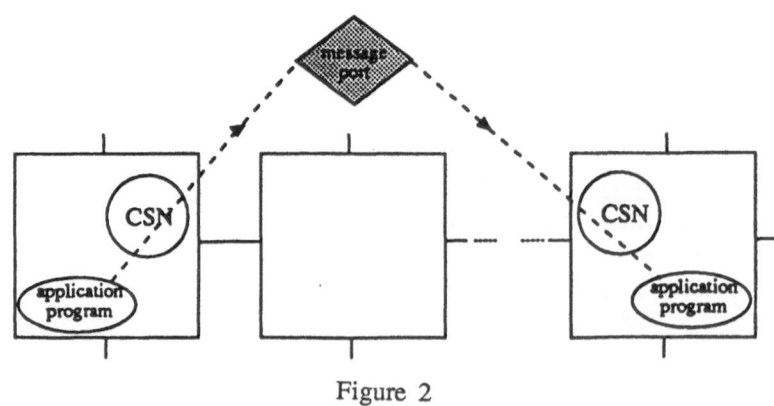

Figure 2

facilities to user application, such as file I/O, printing on the screen and reading from the keyboard. Thus, the full range of standard I/O library calls is available to the user.

The programmer has to describe the structure and layout of the parallel application: which process runs where, how is it connected, ... The simplest approach within CSTools is the PAR file loader, which reads a .par file and loads the application accordingly. An example of such a PAR file is:

```
par
                processor 0 (memory 2m) master.ex8
                processor 1 for 7 slave.ex8
                processor 8 (board mk052,proc_id m1b3) gfx.ex8
                networkis binarytree
                closeto 0 8
end par
```

In this example, processor 0 has at least 2 MB of memory and runs the program master.ex8, processors 1, 2, 3, 4, 5, 6 and 7 run the program slave.ex8, processor 8 must be an MK052 graphics board, module 1, board 3 and the program gfx.ex8 is run on it. Processor 0 (the master) sends the work to the slaves, collects the results and displays them on the neighbouring processor 8. The master processor and the graphics processor should be neighbours and the topology chosen is a binary tree.

As with other parallel programming environments, once a program is working using a simple .par description, it can be tuned by choosing a more suitable network topology, or by moving heavily communicating processes closer together, without changing the source program.

THE EXPRESS PORTABLE PARALLEL PROGRAMMING ENVIRONMENT

What is Express?

Express is a parallel programming environment that allows the user to create parallel C and Fortran programs in a similar way to writing sequential programs. It addresses the needs of parallel programs for message passing, I/O, graphics, data decomposition and load balancing, communication between the parallel machine and a host, configuration management, ... It was developed at the California Institute of Technology, by the research group of Prof. dr. Geoffrey Fox. Caltech is of course one of the birthplaces of distributed memory systems. The excellent book by Geoffrey Fox et al describes thoroughly the possible applications of such distributed memory machines.

Because Express provides a full layer of system software which frees the programmer from the features of the parallel hardware, programs written using Express are portable among all parallel and sequential systems that support Express. The latter is extremely important when the maintenance of parallel programs is considered: you can still run the same program on a normal sequential machine, so you only need to maintain a single version. It has been successfully implemented on transputer systems, the NCUBE and the IPSC hypercubes, a network of SUNs and a number of shared memory machines.

All aspects of a good parallel programming environment: flexibility, scalability, hardware and topology independence, ease of use, efficiency, load balancing, Unix-like I/O, portability, semi-automatic data and process decomposition, distributed graphics, configuration management, ... can be found in Express.

Express also provides some tools for diagnosing and understanding parallel processing problems. It has a source-level debugger and a truly parallel performance monitor. The network debugger NDB brings the power of source-level debuggers found on conventional sequential computers to the parallel world. NDB is a fully parallel source-level debugger (for C and Fortran at the moment) based on Unix' dbx. The performance monitoring tool PM allows you to detect communication bottlenecks and load imbalance problems in a graphical way.

Summarized, the Express operating environment contains the following tools:
- the configuration management tool CNFTOOL for setting up and modifying the network topology
- high-level and low-level communication primitives for sending messages to other transputers, hosts and peripherals
- a transparent I/O system, that gives every node access to operating system facilities
- a multitasking system, based on communication: the reception of a particular message starts off a task on that processor
- a parallel graphics system, that makes device independent graphics available to all processors
- the network debugger (dbx-like) NDB for debugging parallel programs at the source level
- the performance monitor PM, for evaluating, analyzing and enhancing the performance of a parallel program.

Why Express?

Express is based on a number of ideas:
- it is unreasonable to rewrite 30000 lines of CFD code (originally written in Fortran) in some clever and new parallel language (e.g. Occam) and debug it. Express is a language extension to C and Fortran to make the porting process easier.
- it is easier to work in a familiar operating environment (even if it is MS-DOS) than with a totally new operating system. Express provides services to run applications on a parallel computer, either shared memory or distributed memory machine (e.g. transputer) and it can happily coexist with operating systems as Helios and MeikOS on the transputer or MS-DOS, VMS and Unix on the host. In a sense, it plays the role of an operating system for the parallel system, as it provides the ability to communicate with other processors and hosts, share data, read files, do graphics, debug applications, analyze performance, etc. The user interface (to list directories, print files, edit, ...) are still dealt with by the familiar operating system.
- a program should be able to run on 4 as well as 16 processors, by providing a parameter at runtime and without changing and recompiling the actual program. Express provides a number of functions for doing this at runtime.

SOME FEATURES OF EXPRESS

Configuration

Transputer-based systems have the interesting capability of being reconfigurable, either manually or electronically, to match the topology to a particular application. Express hides most of the hardware and interconnections from the user, but in order to tell Express how the underlying topology looks like, the user must run CNFTOOL to set up the configuration. From then on Express will automatically forward messages regardless of the connectivity of the machine. Of course, certain algorithms will give a better performance if a particular topology is chosen.

As a result, CNFTOOL will be run on two occasions:
to set up the initial system and to optimize the configuration for a particular application.

Once a system is correctly configured using CNFTOOL, it is necessary to start the Express kernel. The command "exinit" resets the network, downloads the kernel to all node processors and starts it running. It also performs any necessary hardware configuration on systems which support this. The command "exstat [-l]" gives a summary of the current system usage. It shows the number of nodes available and the number of nodes in use. The switch "-l" also shows which process is allocated to which node.

Interprocessor Communication

The lowest level of the Express system is based on an asynchronous, point-to-point message passing system. This means that any node can send messages to any other node in the system (including hosts) at any time. The kernel will buffer and route the messages to their destination. At the user level, Express offers both synchronous and asynchronous functionality: normal I/O functions will block, but there are provisions for non-blocking I/O as well.

Messages have a certain length, source and destination and contain of course some data. They also have a message type, which can be used to specify that certain messages are intended for one task rather than another. Wildcards (the value DONTCARE) can be used both for message source and type. Message data can be anything (integers, floating point values, strings, structures, ...), it is a stream of bytes without any structure.

The function

exwrite (buffer, length, &dest, &type)

sends a message "buffer" of the specified "length" to the node "dest" with message type "type". Zero length messages can be used, e.g. for processor synchronization. The host has the special processor number HOST, defined in the include file "express.h. The expression

status = exread (buffer, length, &src, &type)

will wait until a message of the specified type "type" and specified source node "src" is received. A maximum of "length" bytes is transferred in the user supplied "buffer" and any extra bytes are discarded. The number of bytes read is returned in the variable "status". If more than one message has been received, the first one that meets the specifications is read.

If the value DONTCARE is used, then any value is considered a match and the accepted value is copied in the user supplied parameter, either source node or message type. Blocking I/O routines are of course a potential danger for deadlocks. If such a deadlock occurs, the send/receive functions should be the first place to look for the origin of deadlock. Remark that sending messages can proceed (as long as the internal Express buffers are not full), but that the receive operation is blocking.

To see if a particular message is available without actually reading it, a function "extest" is available:

status = extest (&src, &type)

A negative return value status indicates that no suitable message is currently available. Otherwise, the status value indicates the length of the appropriate message waiting for reception. In addition to the wildcard ability, we can also specify ranges of message types, using the "exinctype" (include these types in the DONTCARE specification) and the "exexctype" function (exclude these type when using DONTCARE).

Furthermore, global broadcast operations to some or all nodes ("exbroadcast"), send/receive combinations between two nodes ("exchange"), gather operations from some or all nodes ("excombine"), concatenations of messages ("exconcat"), vector read/write/exchange operations ("exvread", "exvwrite", "exvchange") and send/receive operations ("exwritefd", "exreadfd") to disk are also possible.

Two of these require special attention: the "exchange" and the "excombine" operations. It happens very often in scientific applications that nodes have to exchange data. This involves an exread and exwrite operation by both nodes and doing these in the wrong order may cause a deadlock. To avoid this and to increase performance, the "exchange" function was introduced. For similar reasons (communicating large chunks of data is more efficient than communicating several smaller chunks) the functions "exvread", "exvwrite" and "exvchange" were introduced. The latter exchanges vectors of data between two processors and due to a skip parameter these vectors can be rows, columns or more complicated structures in a matrix. These functions are frequently used in conjunction with the exgrid functions, described below.

The function

excombine (data, function, size, nitems, Nnodes, Nlist)

combines the "nitems" data items, each of size "size" using the function "function". The last two arguments describe which nodes the function should be applied to. Because the order of arrival of the data items is unknown, only functions that respect commutativity and associativity rules (e.g. multiplication, addition, averaging) are allowed.

Non-Blocking Communication Functions

This is implemented with the "exreceive" routine:

exreceive (buffer, length, &src, &type, &status).

Its function is very similar to exread, but it is non-blocking and returns a non-negative value, indicating the length of the message read, in "status" when the message has arrived. If this happens, "src" and "type" reflect the source node and message type as with exread, but the user has to make sure that they are not overwritten (typically by using two of each). This routine can be applied to introduce a double buffering strategy,e.g. to avoid that processors are sitting idle, waiting for data, when they are sending results back to a master.
Similarly, we can minimize the time to read data from disk and do some useful work in the meantime. Another usage can be found in a pipeline, where we want to keep data flowing smoothly through a pipe.

Topology Independent Communication (the exgrid() Library)

In this section we describe a set of routines, often called the exgrid() library, with which a considerable amount of automatic decomposition is possible, which hides the details of the parallel machine from the user. These routines can be applied on a large number of problems, often found in science and engineering (finite element analysis, relaxation methods, image processing, CFD, ...). The result is highly portable code, that can be implemented on both distributed and shared memory machines and also on sequential machines, since it is independent of the underlying processor topology.

The exgrid library functions were introduced to map the user's problem domain, which has the topology of a cartesian grid in N dimensions, onto an unknown hardware topology.

The key structure that describes the runtime environment (the number of processors available and a unique identifier of the processor) has the type "struct nodenv", which can be filled using a call to the "exparam" function:

```
#include "express.h"
{
            struct nodenv nodedata;
            exparam(&nodedata);
}
```

Basically, the user tells the system which way the application level data is distributed (one-, two-, three-dimensional) and Express does the rest. The user does not need to know the exact location of the processors on the machine or the nodes they have to communicate with, it is all handled transparently. The data space is subdivided and assigned to the available processors at runtime and the communication with neighbouring processors from the users viewpoint is extremely easy. Once the number of available processors is determined using "exparam", we can divide them up depending on the application. We can also let Express do this using the "exgridsplit" function, which splits the number of processors in the specified number of dimensions in an even way. At this moment Express knows the dimension of the problem we need to solve and the number of processors in each direction.

If we want to know the coordinates of a particular processor in the physical grid of processors, we use

exgridcoord (processnumber, proccoord),

where "proccoord" is an array of the problem dimension where the coordinates of the processor "processnumber" are returned in (starting from 0).

Of course we will need to communicate with other processors and hence we will have to know their processor number to use it in the exread, exwrite and exchange calls. It will not be a surprise that Express offers a function for doing this. The "exgridnode" routine returns the process number of the processor in a particular direction, that is a specified number of hops away (normally 1 or -1 to access the neighbours). This raises the question of boundary conditions. By default exgridnode assumes that the user domain is periodic and wraps around at the edges. This is useful for a number of scientific applications with periodic behaviour, but it is not generally suitable. In this case, we can override the default behaviour by the "exgridbc" call, which will denote that no communication should be attempted at the edges. All this is handled completely inside Express and the user should not be concerned with this after using the exgridbc call. Because this is all done internally, the same code will run on a sequential computer, which has obviously no neighbours in either direction.

The "exgridsize" function is used to distribute an array over the user grid, and returns the sizes of the subregions assigned to each processor, as well as the index array of the first element stored in this processor. It will often be necessary to send, receive or exchange arrays of data. The routines exvread, exvwrite and exvchange do exactly this and the vectors do not to be contiguous memory locations.

Cubix

Using most environments, parallel programs for parallel computers consist of two parts: a master process running on the host and a server running on the parallel machine. Of course, this can be done with Express as well, but you will see that this host server will always have nearly the same functionality. Cubix offers a different approach. Once the program is loaded into the nodes, that program takes over the control of the machine. The host process is no more than a file and I/O server, but it does not have to be changed from one application to the next. This offers some serious advantages, although it will not always be possible to use this approach.   Program development is easier, because we don't need to write and debug a separate program for the host. Parallel programs will also use standard I/O routines, rather than machine dependent system calls, which make the writing and debugging of those programs simpler.   Furthermore, portability is easier. Many programs have been written which can be compiled and run on a sequential machine, as well as a parallel machine running Cubix and Express. A disadvantage of Cubix seems to be the increased code size of node programs, because all computation, normally done by the host, is now done in all nodes. It is not slower, but it increases the code size. A version of Cubix has been running at Caltech since 1986.

Cubix is a full-function I/O and operating server that enables distributed applications, running on the nodes, full access to the host's resources. It offers three totally different modes: synchronous mode (all processors make requests together and all receive the same response), multiple mode   (all processors make requests together and each receives a different response) and asynchronous mode (any processor can make a request at any time and each is serviced independently). The I/O mode is specified for each stream independently and applications can switch between them.

Using a universal host program, programmers only write one program for any application, accessing the host system using standard I/O functions printf, scanf, getc, ... Of course, to support parallel computers a number of functions were added for doing synchronous and asynchronous I/O, for handling multiple hosts, ... Synchronous programs show a uniform behaviour from processor to processor: they calculate for some time and then output data to the host or to each other. A large number of applications in science and engineering fall into the synchronous category.

Plotix

Plotix is a graphical subroutine library designed for parallel machines. It produces device-independent output for a wide variety of different devices (Tektronix 4010 and 4105, HPGL, EGA, Halo, Sun Windows, Postscript devices, ...). It can be used for drawing pictures, generating menus and for showing what is going on inside the parallel system (the performance monitoring tool PM is based on Plotix). It provides a simple and low-level graphics interface and a few high-level functions such as contouring and clipping. Nodes can be drawing the same image (e.g. a menu bar), drawing individual parts of the image simultaneously (updating the screen at the same time) or drawing parts of the screen independently. Plotix is also available for sequential computers, enhancing program development and portability.

AN EXAMPLE: TRANSPUTER IMPLEMENTATION OF THE KOHONEN FEATURE MAP

Neural networks are based on the structure and functionality of the human brain. These networks are solving problems that a five-year-old can do easily, but these problems are difficult or impossible to program. Neural networks present a major breakthrough in computer science, with a number of advantages and a wide range of applications. Recently, neural networks were introduced in a number of industrial applications. During the learning phase they are trained to associate identifications with examples and they memorize them by altering values in their characteristic. Once trained, they recognize and classify an example pattern more quickly than any algorithmic program would do. Pattern recognition is an ideal domain for applying neural nets. These applications require very expensive high-speed computers to run in real-time and neural networks provide a far simpler and more powerful solution. Of course the performance bootleneck is still there, but it has been moved from the moment of application to the learning phase. They are more powerful, because they can recognize (as humans do) partially incorrect or incomplete input. Neural networks can be viewed as associative memory, associating input patterns with desired output patterns. The time required to do this is very small and independent of the number of associations stored. Some companies are implementing neural nets in silicon, but the only cost-effective way to use neural networks at the moment is still by simulating them on a conventional computer. The learning phase is the performance bottleneck and can be very computation-intensive. Therefore, a lot of researchers are examining the use of parallel computers to implement these neural net simulators.

The problem presented here is the classification of input patterns by training a neural network using Kohonen's feature map learning algorithm. The algorithm produces a mapping from an n-dimensional space onto a two-dimensional grid of neuron-like elements. Associated with each neuron are a set of synapses and a set of pre-synaptic neurons that constitute an input layer. Initially all synapses have random weights. The synapses are adapted according to the best match between the activation of the input neurons and the synapses. Adaption occurs by slightly changing the weights of the most similar synaptic vector and its neighbours, so that the similarity of that synaptic vector with the input vector increases. More information about this application can be found in [Leman 1989].

The algorithm can be summarized as follows:

(1) Initialize the synaptic weights by giving them a random value
(2) Initialize the learning rate and radius of the neighbourhood
(3) Randomize the input data
(4) Compute the Euclidian distance between the input vector and the synaptic vector
(5) Find the neuron whose synaptic vector is most similar to the input vector
(6) For all neurons within the neighbourhood radius of the neuron chosen, adapt the synapses according to the formula:

$$w(t+1) = w(t) + a(t) * (v(t) - w(t))$$

where $w(t+1)$ is the synaptic vector at time $t+1$, $w(t)$ the synaptic vector at time t, $a(t)$ the learning rate at time t and $v(t)$ the input vector at time t.

(7) Decrease the learning rate and neighbourhood radius if necessary
(8) Go back to (3).

The implementation of this algorithm on an arbitrary transputer network is very straightforward. The following program shows the framework of the implementation using Express:

```c
#include <stdio.h>
#include <express.h>

main ()
{
        /* determine the configuration at run-time */
        exparam (&nodedata);

        /* split the processors in 2 dimension */
        exgridsplit (nodedata.nprocs, 2, nprocs);

        /* initialize Express to work with a 2D user space*/
        exgridinit (2, nprocs);

        /* find the piece of the 2D data space, assigned to this processor */
        exgridsize (nodedata.procnum, neurogrid, my_neurogrid,
                my_neurogrid_start);

        ... initialize synaptic weigths (random value)
        ... initialize data structures

        /* if graphics is wanted, open display on every processor */

        if (display_matrix)
                openpl (32000, (FILE *) 0);

        for (all iterations) {

                ... randomize the input data

                for (all input vectors) {
                        ... find most similar vector on each processor
                        ... find the most similar vector on all processors (excombine)
                        ... distribute this information to all processors (exconcat)
                        ... adapt the synapses in the neighbourhood of this neuron, inhibit
                        the others
                        ... if (display_matrix && (vector_number == display_vector))
                        ... display_matrix[i][j]
                }
        }
        if (display_matrix)
                close_pl ();
}
```

The resulting parallel program, which can be run on a single transputer system using:

        cubix neuron

and on 4 transputers using:

        cubix -n4 neuron,

looks very much like a standard C program. The input/output, initialization and dynamic memory allocation sections remain exactly the same. In fact, a comparison between the sequential and parallel version shows the major advantage of Express for this application: we were able to parallelize the program by adding a very small number of Express library calls. Not all applications will be as easily parallelized as this, but many other problems in science and engineering fall into the geometric and data parallel class. In fact, it should be possible to do the conversion of a sequential application of this kind to a parallel version automatically, and we are working on this as part of the ESPRIT Parallel Computing Action. We will need real test cases and we invite companies with such applications (either Fortran or C) to come and talk about a possible collaboration to convert their code.

For a test of 20 iterations with 115 input vectors of length 12, the following results were obtained:

	10 x 10 grid		40 x 40 grid	
	with	without	with	without
	display		display	
1 processor:	4'00"	1'50"	???	25'40"
4 processors:	3'00"	45"	20'32"	6'47"
Speedup:	1.3	2.44	???	3.8

For large networks of neurons, the execution time increases dramatically. We can see that the simulation on 1 T800-20 processor takes nearly half an hour. On a simple PC/AT, the program takes approximately 20 times more. Popular processor combinations, such as Intel's 80386/80387 or Motorola's 68030/68882 will need between half an hour and one hour. Thus, the performance of the T800 processor is already quite spectacular, but a 4 processor system runs nearly 4 times as fast. Unfortunately, the time to display the data every iteration is enormous and slows down the execution considerably, although the ability to do graphics and textual output from every node was found to be very useful.

The parallel version of the program did not work first time (remember that this was our first Express program and the documentation is far from excellent). The associated network debugger NDB closely resembles the standard Unix debugger dbx, but still needs some debugging itself.

Another problem came from the fact that we use Parsytec boards, which are not fully compatible with other boards. An intermediate solution was found, but we keep on working actively together with Parasoft to solve this and other problems with Express.

Once the program is working, it can be optimized using the performance monitoring tools, which immediately showed that parallel graphics is very slow. The topology of the network can be modified to improve the performance of the program.

An interesting aspect of this program conversion was to see how well current transputer hardware and software is adapted to transputer outsiders. This is extremely important for the wide-spread use acceptance of parallel processing by non-computer science experts. Although this is a study in its own right and could be the subject of another paper, the conclusion would probably be that transputer technology is not yet ready to be used by everyone. The step from a sequential program to a program which runs on more than one processor is big and not so easy with the tools of today (but highly application dependent), but Express is certainly one of the better ones. What we really need is a well-defined (and well-documented) virtual machine interface which frees the programmer from the actual hardware (topology, number of nodes, type of processors, ...) of the machine he or she uses.

The main advantages of Express

- Users of parallel systems can develop, debug and offer finished applications in a familiar environment. There is no need to abandon your common, well-acquainted operating system to run applications on the parallel computer.
- Express supports several programming paradigms: applications may run completely on the parallel computer or part of the code may be farmed out to the parallel machine, while other parts still run on the host computer.
- Powerful semi-automatic decomposition tools allow many scientific applications to be parallelized with little effort.
- It supports both static and dynamic load balancing of data decomposed problems.
- Programs can be made scalable: to take advantage of more processors, just change a command-line parameter. There is no need for recompiling or reconfiguring the program.
- It is portable: the same program can be run on transputer systems, NCUBE, Intel Hypercube, on a network of personal computers or workstations or even on a sequential machine.
- It was designed by users of parallel machines, who incorporated the things they needed most and it relies on 5 years of parallel processing research at the California Institute of Technology.
- Both NDB and PM are very powerful tools for debugging and optimizing parallel programs.
- Programs written using Express are independent of the topology of the underlying transputer system. Optimization of the network can be done after the program is debugged and running.
- Express is not perfect, but it is a big step in the right direction.

A COMPARISON OF EXPRESS AND CSTOOLS

To show the basic difference between Express and CSTools, we need to say a little bit about the types of parallelism that can be introduced into an application: algorithmic parallelism, geometric parallelism and farming.

Algorithmic parallelism means the division of the application into modules, executed concurrently on multiple processors. It is fairly easy to implement, because there is no need to modify much of the existing program. If the application was written with modularity and data abstraction in mind (but not may dusty deck programs are), this is a lot easier.

For geometric parallelism the application is split into independent data sets, according to the geometry of the problem. Twodimensional and threedimensional decompositions of the data domain are very popular and can be used for various applications. Each processor works on a fixed piece of the data domain, but the program code remains roughly the same. Introducing geometric parallelism is obviously somewhat more difficult than algorithmic parallelism, but the efficiency can be very high. Of course this type of parallelism works best if the amount of work in each part of the geometry is large and approximately the same.

Farming is similar to geometric parallelism, but applies also to applications where the amount of work varies with the complexity of the problem (e.g. mandelbrot pictures). The input data is also partitioned into independent sets of data and distributed over a transputer network. The topology of the processor farm is independent of the application geometry: work is sent into the network of processors, done by a free processor and results are sent back to the master processor. Load balancing is dynamic and the efficiently can be very close to 100 %.

In fact, succesful use of massively parallel machines at the moment are found in applications with either geometric parallelism or farming (often jointly called data parallel applications). The data set can be a neural network, a database, the possible moves in a chess program, an image, a matrix, the 3D space in a weather forecast program, ... Many applications in physics, biology, chemistry, engineering, ... fall into these classes of parallelism and are tackled rather easily and very efficiently by massively parallel computers.

Table 1

	Express	CSTools
Implemented on:	Sun, transputers (any manufacturer, IPSC, NCUBE, Symult, ...	Meiko, Sun
Configuration tools:	cnftool	electronically reconfigurable
Mapping:	host-node approach, exopen()	PAR file loader
Message passing:	processor-to-processor	process-to-process, named message ports, Transport, CSN
Runtime development tools:		
- symbolic debugger	NDB (based on dbx)	tdb (based on dbx) tdbtool (based on Sun's dbxtool)
- performance monitor	PM	
Types of parallelism:		
- Algorithmic:	using exopen(), ...	PAR file loader
- Geometric:	exgrid() library	using cs_getInfo(), ...
- Farming:	exgrid() library	master + multiple slaves

Examples of such applications are:
- Monte-Carlo simulation
- Finite element applications: structural analysis, fluid flow, electromagnetics, ...
- Astronomical data analysis
- Quantum dynamics
- Plasma phsics
- Computer chess
- Neural network simulation
- Ray tracing and computer graphics
- Kalman filters
- Aerodynamics
- Image processing
- Computer graphics

Express is obviously highly suited for data parallel applications, whilst CSTools is more aimed at algorithmic parallelism. Although there is a clear difference between the two parallel programming environments discussed, with some extra efforts, Express can be applied to algorithmic parallelism and CSTools can be used for geometric and data parallelism. In fact, the basic functionality provided can be used to implement the exgrid() library under CSTools or the PAR file loader and process-to-process communication under Express. It would be a good thing if Meiko and Parasoft could agree to take the best of both worlds and integrate it into a new, standard parallel programming environment. A comparison is given in Table 1.

CONCLUSION

We don't think that massively parallel computers will ever be able to replace the traditional supercomputer, but they are definitely a very useful and very powerful addition to the computing environment of the modern researcher. The transputer is the ideal component for building such massively parallel machines and has given a boost to parallel computing. But there is a problem: can your software cope with this challenge? Will you have to make massive investments in rewriting already written sequential code into a strange parallel language to stay in the business? We need new tools to solve this problem, either new programming languages such as Strand88 or Linda, new operating systems such as Helios or Mach, or new parallel operating environments such as CSTools or Express. We need a solution that works for lots of programmers and lots of applications on lots of machines. The three major requirements of these tools are believed to be ease of use, portability and efficiency. We opt for a parallel programming environment, which is machine and configuration independent, which makes it easy to integrate existing software and which is easy to learn and efficient. The two operating environments discussed, CSTools and Express, have a lot of similarities and a lot of differences. It would be wonderful if we could take the best of both worlds and build a new, standard, portable parallel programming environment. We are very positive about Express, which is used on a number of PC plug-in boards. We hope to be able to say the same about CSTools on a Meiko In-Sun Computing Surface, which is to be delivered real soon now.

Once there is a high-level software environment for communication, synchronization, program decomposition and data decomposition, parallel programming is almost as simple as sequential programming. Of course, the more the environment does for you, the better. Efficient library implementations (e.g. BLAS) are equally important, but one of the basic obstacles, the development of parallel algorithms, can only be solved by adjusted education.

Parallel distributed memory computers will become extremely powerful (e.g. with the Inmos H1 next-generation transputer) and very efficient (with through-routing done in hardware). They are already very cost-effective for a large number of applications. Parallel programming environments such as Express and CSTools will make parallel programming simple, portable, efficient and accessible to everyone.

BIBLIOGRAPHY

Parallel Processing

Fox, G., Johnson, M., Lyzenga, G., Otto, S., Salmon, J., and Walker, D., *Solving Problems on Concurrent Processors*, Volume 1, Prentice-Hall (1988).

Gustavson, J.L., Montry, G.R., and Benner, R.E., "Development of parallel methods for a 1024-processor hypercube", *SIAM J. Sci. Stat. Comp.*, no. 9, 609-638 (July 1988).

Transputers

Hamilton, A., "Some Issues in Scientific-Language Application Porting and Farming using Transputers", Inmos Technical Note 53 (July 1989).

Stiles, G.S., "How the transputer stacks up to other processors: a comparison of performance on several application programs", *Proceedings of the North-American Transputer Users Group, Salt Lake City, April 5-6, 1989.*

Van Renterghem, P., "De Transputer: sleutel tot de toekomst", *Technology Transfer Express*, 30-36 (June 1988).

Van Renterghem, P., "Transputers - De toekomst ziet er parallel uit", Technisch Management (May 1989).

Van Renterghem, P., "Transputers for industrial applications, concurrency: practice and experience", *Wiley Journals* (to appear).

CSTools

Meiko Ltd., CSTools Documentation Guide (1989).

Express

Flower, J., Kolawa, A., and Liang, T., "Expressing Cosmos", *Proceedings of the Liverpool Conference, Liverpool, Aug. 23-25, 1989.*

Van Renterghem, P., "Express: a parallel programming environment", *Proceedings of the ISPRA Course 'Parallel Processing with Personal Computers, 5-9 June 1989.*

Neural Networks

Auger, J.M., and Dorizzi, B., "Implementation de l'algorithme de cartes topologiques de Kohonen sur Transputer", *La Lettre de Transputer*, 17-29 (September 1989).

Kohonen, T., *Self-Organization and Associative Memory*, Springer-Verlag, Berlin (1984).

Kohonen, T., "The 'neural' phonetic typewriter", *IEEE Computer*, 21 (3), 11-22 (1988).

Leman, M., and Van Renterghem, P., "Transputer implementation of the Kohonen feature map for a music recognition task", *Proceedings of the Second International Transputer Conference, BIRA, Antwerp 23-24, 1989.*

Lippmann, R.P., "An introduction to computing with neural nets", *IEEE ASSP Magazine*, 4-22 (April 1987).

Linda (reprints available from Yale University)

Carriero, N., and Gelernter, D., "Applications experience with Linda", *Proceedings of the ACM/SIGPLAN PPEALS* (1988).
Carriero, N., and Gelernter, D., "How to write parallel programs", *ACM Computing Surveys* (November 1988).
Carriero, N., and Gelernter, D., "Linda in Context", *Communications of the ACM*, Volume 32, No. 4 (April 1988).
Gelernter, D., "Getting the job done", *Byte* (November 1988).

Helios

Perihelion Software Ltd., *Helios Technical Notes 1-21*, Perihelion (1989).
Perihelion Software Ltd., *The Helios Operating System*, Prentice Hall (1989).

MONTE CARLO METHODS IN CLASSICAL STATISTICAL MECHANICS*

Martin Schoen
Institut für Experimentalphysik, Naturwissenschaftliche Fakultät, Universität Witten/Herdecke, Stockumer Str. 10, 5810 Witten, F.R.G.

Dennis J. Diestler
Richard B. Wetherill Laboratory of Chemistry, Purdue University, West Lafayette, IN 47907, U.S.A.

John H. Cushman
Lilly Hall of Life Sciences, Purdue University, West Lafayette, IN 47907, U.S.A.

I. INTRODUCTION

I.1. General remarks

The introduction of statistical concepts into physics during the second half of the nineteenth century was a major step towards a better understanding of the behavior and properties of matter. Around 1850 the chemist August Karl Krönig [1] reintroduced the idea of equivalence between heat and atomic motion which in the following years motivated the pioneering work by Clausius and Maxwell culminating in the virial equation of state for real gases [2] and the first realistic theory of transport processes in dilute gases [3]. It was then at the turn of the century when Ludwig Boltzmann [4] formulated his H-theorem, which expresses a macroscopically introduced entropy in terms of microscopic variables such as particle positions $\mathbf{r}$ and momenta $\mathbf{p}$ via a non-equilibrium distribution function $f(\mathbf{r}^N, \mathbf{p}^N, t)$, where the superscript indicates that in general f depends on the entire set of coordinates $\{\mathbf{r}\}$ and momenta $\{\mathbf{p}\}$ of all N particles; t denotes time. This finally closed the gap between classical phenomenological thermodynamics and an atomistic description of matter.

All measurable quantities of any system can thus be expressed as averages of suitable microscopic variables over $f(\mathbf{r}^N, \mathbf{p}^N, t) = f(\Gamma^N(t))$, where $\Gamma^N(t)$ denotes a point in 6N-dimensional phase space (3N coordinates and 3N momenta). A succession of

---

* Supported in part by the United States Department of Energy Grant DE-FG02-85 ER 60310.

points $\Gamma^N(t_0)$, $\Gamma^N(t_1)$, ..., $\Gamma^N(t_n)$ with $t_0 < t_1 < ...... < t_n$ defines a trajectory in phase space. However, the analytical calculation of a phase space trajectory is a highly non-trivial and practically insoluble problem because of the enormous number of variables involved (macroscopic piece of matter: $N \approx 10^{23}$). Additional model-dependent assumptions have to be introduced in order to solve Liouville's equation, which governs the time evolution of the orbit $\{\Gamma^N\}$ as far as classical systems are concerned [5].

I.2 Low Density Systems

Consider, for instance, a very dilute gas of atoms for which the mean free path is fairly large. These particles move freely most of the time, but every now and then collisions between them occur. Since most of the space occupied by the particles is empty, it is reasonable to assume that only two particles collide at a time. Under this assumption it can be shown that $f(\mathbf{r}^2, \mathbf{p}^2, t)$ suffices to describe the time evolution of the system in phase space. This reduced two particle distribution function may be obtained as a solution of Boltzmann's equation, which can be derived from the more general Liouville equation [6] via an equation of lowest order in the so-called BBGKY (Born-Bogolyubov-Green-Kirkwood-Yvon) hierarchy of integro-differential equations. In the BBGKY hierarchy the unknown distribution function of lowest order, $f^{(1)} = f(\mathbf{r}^1, \mathbf{p}^1, t)$, is expressed in terms of the likewise unknown two particle distribution function $f^{(2)}$. An additional relation between successive distribution functions is therefore required at each level of the hierarchy to solve any of the BBGKY equations. These additional equations are known as closure relations. In the case of Boltzmann's equation such a closure relation is introduced via the concept of molecular chaos, namely $f^{(2)} = f(\mathbf{r}_1, \mathbf{p}_1, t)f(\mathbf{r}_2, \mathbf{p}_2, t)$, *before* and *after* the collision which in effect ignores all correlations between particles except for a very small time interval t during which the collision takes place. The subscripts on $\mathbf{r}$ and $\mathbf{p}$ label the two particles in question.

At higher, say liquid densities, such a description of phase space evolution is no longer appropriate because characteristic mean free paths are now comparable to the range of intermolecular interactions. Particles are not interacting via successive, uncorrelated binary collisions but are more or less in continuous contact with nearest neighbors. Thus the assumption of molecular chaos is invalid; in fact, one has to consider correlations between particles via higher order distribution functions.

I.3 Dense Fluids

Consider, for example, a dense fluid in thermodynamic equilibrium. $f^{(n)}$ can then be replaced by the equilibrium distribution function $f_0^{(n)}$ $(\mathbf{r}^n, \mathbf{p}^n)$ which no longer explicitly depends on time. Furthermore, $f_0^{(n)}$ can be factorized as $f_0^{(n)}$ $(\mathbf{r}^n, \mathbf{p}^n) = P^{(n)}$ $(\mathbf{p}^n, \rho^n)$ where $P^{(n)}$ is a product of independent Maxwell-Boltzmann distributions and $\rho^{(n)}$ is the n-particle equilibrium density [6]. Starting from the BBGKY hierarchy with these assumptions, we eventually arrive at a similar hierarchy for the $\rho^{(n)}$ known as the hierarchy of BGY (Born-Green-Yvon) integro-differential equations [6], which form the basis of the equilibrium theory of dense fluids.

If we restrict ourselves to a fluid with pairwise additive interactions, which is an excellent approximation for the overwhelming number of systems, $\rho^{(2)}$ turns out to be the most important one of the n-particle densities for two reasons. First of all, $\rho^{(2)}$ is related to a pair correlation function $g^{(2)}(\mathbf{r}^2) = \rho^{(2)}(\mathbf{r}_1, \mathbf{r}_2) / \rho^{(1)}(\mathbf{r}_1) \rho^{(1)}(\mathbf{r}_2)$ which can be measured in x-ray scattering experiments [7]. Secondly, all thermodynamic functions, such as internal potential energy U and pressure P can be written as integrals over $g^{(2)}$.

However, $g^{(2)}$ is theoretically difficult to obtain for the following reasons. Starting from the appropriate equation in the BGY hierarchy, one finally obtains another integro-differential equation relating $g^{(2)}$ to the triplet correlation function $g^{(3)}(r_1, r_2, r_3)$ for which an additional, a priori uncontrollable approximation has to be made. Amongst several others the most famous such approximation is the one introduced in 1935 by Kirkwood, who replace $g^{(3)}$ by a triple product of pair correlation functions, namely $g^{(2)}(r_1, r_2)\, g^{(2)}(r_2, r_3)\, g^{(2)}(r_1, r_3)$ [8].

Unfortunately, this closure relation, the so-called superposition approximation, fails at reasonably large densities [9]. More realistic closure relations, on the other h and, immediately lead to much more complicated, sometimes tensorial integro-different ial equations [10], for which no solution has been obtained so far.

I.4. Computer Simulation

Since most of the analytically obtained results in liquid state theory have proved to be only qualitatively satisfactory , computer simulation methods have gained more and more importance during the last few years. In this approach a sequence of particle configurations is generated numerically. Any property of interest is then evaluated as an average over this succession of configurations. There are two different ways of generating such a sequence of particle configurations. In a molecular dynamics (MD) calculation the classical equations of motion are integrated for a certain period of time while in a Monte Carlo (MC) calculation particle configurations are generated by means of a stochastic process, time not being involved explicitly [11]. Therefore, in MD time-dependent properties are accessible in addition to structural and thermodynamic ones, whereas in MC only the latter two can be obtained. What may seem to be a disadvantage of MC at first sight is, however, recouped because algorithms can be designed quite easily to allow for the generation of particle configurations that are compatible with any given statistical mechanical ensemble. In MD, on the other h and, it is rather difficult, if not at all impossible, to derive proper equations of motion for the same purpose. This is especially true for the grand canonical ensemble, where the particle number fluctuates with time . The creation and destruction of particles make it difficult to integrate the equations of motion numerically because of the discontinuous change in interparticle forces whenever the particle number changes. However, as we shall demonstrate later in this article, the grand canonical ensemble may be the most appropriate one to study certain phenomena in statistical mechanics.

In addition, MC calculations can easily be applied to quantum mechanical systems [12]. In MD it is rather difficult to study even semiclassical systems, while a full quantum mechanical treatment of a many particle system has not yet been presented. The first MC [13] and MD [14] calculations in the canonical microcanonical ensemble, respectively, were performed in the late 1950's for idealized fluids composed of hard spheres. Since then computational power has increased astronomically. The complexity and the size of model systems being investigated has grown on a similar scale. Subjects that are under current investigation range from, say, properties of biopolymers near membranes [15] over liquid crystals [16] to solid state physics, where melting has been studied in systems containing as many as 161604 individual argon atoms [17].

Since about 1980 a large and steadily growing body of research in the field of statistical mechanics has also been devoted to properties of liquids near or interfaces [18,19]. This latter group of systems is of great relevance to a variety of problems ranging from, say, tribology to contamination of soils by pollutants. However, the computer time required to study realistic models of such systems can be enormous, even on the fastest vector or parallel computers. In a later section we illustrate the computational demands by a MC study of liquid-gas phase transitions in micropores.

Since computer time is precious, special care has to be taken to design efficient program structures and algorithms. Therefore, we focus on basic program structures that run efficiently on vector machines. Parallel computer architectures are becoming increasingly important, especially because of a price/performance ratio, which is orders of magnitude below that for large vector machines such as the CYBER 200 series (i.e. ETA 10) or CRAY computers. At the same time computational speeds of parallel, so-called transputer systems are large enough to allow for most MC or MD applications in reasonable time spans. We will present some information about statistical mechanical calculations on transputer systems in this article as well.

II. MONTE CARLO CALCULATIONS

II.1. A Simple Example

Since we do not intend to discuss time dependent properties, we will focus entirely on the application of MC to classical statistical mechanical systems in the remainder of this article. The name "Monte Carlo" was first introduced by Metropolis and Ulam in 1949 [20] because of the extensive use of random numbers made in a calculation of neutron diffusion in fissionable material. Generally speaking, MC is a method for evaluating integrals by a sort of hit and miss experiment.

This may be illustrated by a simple example taken from the excellent and comprehensive book on computer simulation of liquids by Allen and Tildesley [11]. Suppose we want to estimate the number π by a hit and miss experiment illustrated in Figure 1: a circle of radius 1 inscribed in a square, both with their centers at O. A number of trial shots are now generated in the square between OABC, each shot being characterized by two random numbers ξ_1 and ξ_2 from a uniform distribution on (0,1). ξ_1 and ξ_2 define a randomly chosen point in the x,y plane. The distance r_ξ of this point from the origin determines whether the shot has landed in the shaded area (r < 1) and a hit has to be scored. Clearly, the probability of a hit is proportional to the area under the curve CA

$$A_{CA} = \frac{1}{4} \pi r^2 = \frac{\pi}{4} \tag{1}$$

Thus, the fraction of hits is given by

$$\lim_{\tau \to \infty} \frac{\tau_{Hit}}{\tau} = \lim_{\tau \to \infty} \frac{\pi/4}{\tau} \; ; \; \pi = 4 \lim_{\tau \to \infty} (\tau_{Hit} / \tau) \tag{2}$$

where t is the total number of trial shots. Typically, π obtained by this prescription is correct to four figures after 10^7 trials (i.e. shots).

Today there exists a wealth of literature on MC methods, their mathematical foundations as well as their application to a variety of problems in various fields [21-23]. Therefore we restrict our discussion in the following paragraphs to the most

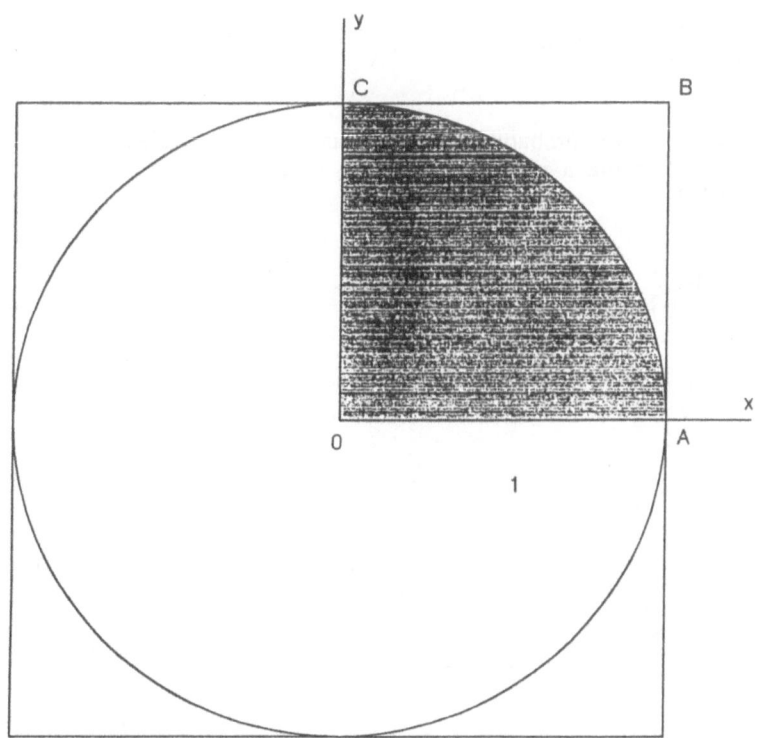

Figure 1. A circle of radius 1 inscribed in a square for the estimation of π by a hit and miss experiment (cf. Figure 4.1 in [11]).

fundamental methodological aspects. For more details we refer the interested reader again to Allen and Tildesley's book [11].

II.2. Outline Of Fundamental Aspects

In the previous paragraph we introduced the MC technique as a general purpose integration technique and used as an illustrative example the evaluation of p, which can be expressed as

$$I = \int_0^1 f(x) \, dx \tag{3}$$

where $f(x) = (1 - x^2)^{1/2}$ denotes the equation of the circle in the first quadrant. Eq. (3) can be rewritten as

$$I = \int_0^1 \left(\frac{f(x)}{\rho(x)} \right) \rho(x) \, dx \tag{4}$$

where $\rho(x)$ is an arbitrary probability density function. If τ trials are performed each one consisting of choosing a random number ξ_τ from $\rho(x)$ in the range $(0,1)$, we may estimate I from [11]

$$I = \lim \left< \frac{f(\xi_\tau)}{\rho(\xi_\tau)} \right>_{\text{trials}} \tag{5}$$

where the angular brackets indicate the average over all trials weighted by $\rho(x)$. In particular, if $\rho(x)$ is chosen to be uniform, that is

$$\rho(x) = 1 \tag{6}$$

we finally arrive at

$$I = \lim_{\tau \to \infty} \frac{1}{\tau} \sum_{i=1}^{\tau} f(\xi_i) \tag{7}$$

While MC is hardly worthwhile for such simple one-dimensional integrands like $f(x)$ in our example, compared with other straightforward numerical techniques (e.g. Simpson's formula) [11], this is not the case for more complex multi-dimensional integrals such as the canonical ensemble partition function

$$Q_{NVT} = \frac{1}{N! h^{3N}} \int \exp[-\beta H(\mathbf{r}^N, \mathbf{p}^N)] \, d\mathbf{r}^N \, d\mathbf{p}^N \tag{8}$$

where $H(\mathbf{r}^N, \mathbf{p}^N)$ is the hamiltonian, $\beta = 1/k_B T$, k_B being Boltzmann's constant and T temperature. In the thermodynamic limit ($N \to \infty$, $V \to \infty$, $N/V = $ const.; N particle number, V volume) the integral in eq. (8) extends over some 10^{23} variables. However, in practice we have to restrict ourselves to much smaller systems containing only a few hundred particles (i.e. integration variables) but even then the dimensionality of the integral in eq. (8) is much too large to permit its evaluation by any method other than MC (or MD (see I.4)).

For the canonical ensemble, in which a thermodynamic state is characterized by a fixed N, V and T, Q_{NVT} is a key quantity. From its relation to the Helmholtz free energy A

$$A = -k_B T \ln Q_{NVT} \tag{9}$$

all other statistical mechanical expressions for thermodynamic quantities can be derived [7].

The integration over momenta can be performed explicitly to give $Q_{NVT} = Z_{NVT}/N!\Lambda^{3N}$, where the so-called configurational integral

$$Z_{NVT} = \int \exp[-\beta U(\mathbf{r}^N)]\ d\mathbf{r}^N \tag{10}$$

can be calculated from

$$Z_{NVT} = V^N \lim_{\tau \to \infty} \frac{1}{\tau} \sum_{i=1}^{\tau} \exp[-\beta U(\mathbf{r}_i^N)] \tag{11}$$

where $\mathbf{r}_i^N$ denotes a randomly chosen point in the 3N-dimensional space of particle coordinates $\mathbf{r}_i^N$.

The evaluation of eq.(11), however, is very problematic in practice. At typical liquid densities the likelihood that two particles overlap significantly for a randomly chosen $\mathbf{r}_i^N$ is quite large. The potential energy $U(\mathbf{r}_i^N)$ for such a configuration is very large and consequently the contribution of the corresponding Boltzmann factor $\exp[-\beta U(\mathbf{r}_i^N)]$ to the sum in eq. (11) is negligible. This will be true for most trials so that an enormous number must be averaged over before the RHS of eq. (11) converges to its "true" equilibrium value for a given thermodynamic state. A similar consideration applies to typical statistical mechanical averages

$$\langle F \rangle_{NVT} = \frac{\displaystyle\sum_{i=1}^{\tau_N} F(\mathbf{r}_i^N) \exp[-\beta U(\mathbf{r}_i^N)]}{\displaystyle\sum_{i=1}^{\tau_N} \exp[-\beta U(\mathbf{r}_i^N)]} \tag{12}$$

if numerator and denominator are to be estimated separately [11]. At this point it should be realized that the MC sampling procedure outlined so far is rather naive because it lacks information about the spatial dependence of $U(\mathbf{r}^N)$, i.e. the significantly contributing Boltzmann factors. A better way to compute statistical mechanical averages of the form given in eq. (12) would emphasize Boltzmann factors that differ significantly from 0. This can be done by a modified sampling prescription known as **importance sampling**, by which configurations are sampled in such a way that the important regions of configuration space are also the most frequently sampled ones [21].

Importance sampling, however, must weight each sampled configuration according to its probability of occurrence. Eq. (12) is then replaced by

$$\langle F \rangle_{NVT} = \frac{\sum\limits_{i=1}^{\tau} F(\mathbf{r}_i^N) \exp[-\beta U(\mathbf{r}_i^N)]/\rho(i)}{\sum\limits_{i=1}^{\tau} \exp[-\beta U(\mathbf{r}_i^N)]/\rho(i)} \tag{13}$$

where $\rho(i)$ is the probability of choosing configuration i. In the case of the canonical ensemble the most natural choice for $\rho(i)$ would be the Boltzmann distribution itself

$$\rho(i) = \exp[-\beta U(\mathbf{r}_i^N)] / Z_{NVT} \tag{14}$$

so that the statistical mechanical average $\langle F \rangle_{NVT}$ can be rewritten as a simple sum over $F(\mathbf{r}_i^N)$ calculated for a sequence of t individual configurations i as

$$\langle F \rangle_{NVT} = \frac{1}{\tau} \sum_{i=1}^{\tau} f(\xi_i) \tag{15}$$

We have thus simplified the formulation of the numerical problem considerably by going from eq. (13) to eq. (15). The price to be paid for this numerical simplification, however, is a new conceptual problem: How can we generate a sequence of configurations according to a predefined probability of occurrence such as $\rho(i)$?

To answer this question the generation of configurations may be viewed as a realization of a Markov chain, which is a sequence of random variables subject to the conditions [24]:

i)   The realization of elements of the chain belong to the set $\{\Gamma_1,\Gamma_2,...,\Gamma_m,\Gamma_n,...\}$ called state space
ii) The distribution of a random variable in the chain depends only on the immediately preceding one, i.e. Markov chains are characterized by a one step memory in state space

Because of ii) two successive states Γ_m and Γ_n are linked by a transition probability π_{mn}. The probability densities ρ_m and ρ_n for the occurrence of the two states are also related via

$$\sum_m \rho_m \pi_{mn} = \rho_n \tag{16}$$

Since the rows of π add to 1

$$\sum_n \pi_{mn} = 1 \tag{17}$$

π is called a stochastic matrix [24]. For more details about π and its properties the interested reader is referred to a number of very useful supplementary publications [21-24]. We assume that a realization of a Markov chain satisfying eqs. (16) and (17) as well as to the condition of **microscopic reversibility**

$$\rho_n \, \pi_{mn} = \rho_m \, \pi_{nm} \tag{18}$$

is ergodic (see Sec. III.2). Ergodicity means that every point in state space can eventually be reached from any other point. For a system truly at equilibrium the Markov chain should be ergodic.

It is also important to ensure that transition probabilities do not depend on the a priori unknown partition function, which comes into play in principle as a normalization factor of the probability density functions ρ_m, ρ_n (see eq.(14)). In other words, an algorithm has to be designed such that π_{mn} depends at most on the ratio ρ_m/ρ_n. Probably the most common such algorithm is the one proposed by Metropolis et al. [25] and often referred to as the "asymmetrical solution" (see [11] for other algorithms). In the Metropolis algorithm the transition probabilities for two distinct states m and n are given by

$$\begin{aligned} \pi_{mn} &= \alpha_{mn} & \rho_n \geq \rho_m \\ \pi_{mn} &= \alpha_{mn} \, (\rho_n \, / \, \rho_m) & \rho_n < \rho_m \end{aligned} \Biggr\} \quad m \# n \tag{19}$$

The probability that the system remains in the same state is from eq. (17)

$$\pi_{mm} = 1 - \sum_{n\#m} \pi_m \tag{20}$$

In eq. (19) α is a symmetric stochastic matrix, the so-called underlying matrix of the Markov chain [24].

While MC is not limited to classical systems, for most liquids (except, say, H^2 or He) a quantum mechanical treatment is unnecessary because typical intermolecular distances are large compared to the thermal de Broglie wavelength characteristic of most liquids [7].

III. THE SCHEME IN PRACTICE: MONTE CARLO IN THE CANONICAL ENSEMBLE

III.1. Implementation

To implement the Metropolis algorithm, we need to specify α in such a way that the system will be carried from state m to any of its neighboring states n with equal probability. One possible, though completely arbitrary definition of a neighboring state is illustrated for a two-dimensional system in Figure 2, where six atoms are shown in state m. A neighboring state is now generated in the following way: One of the six atoms, say i, is chosen at random and randomly displaced from its old position $\mathbf{r}_i^m$ at the center of a small square of side length δr_{Max} to a new position $\mathbf{r}_i^m$ still inside the square by means of

$$\mathbf{r}_i^m = \mathbf{r}_i^m + \delta r_{Max}\ (2\xi - 1) \tag{21}$$

where ξ denotes a two-component vector of random numbers on (0,1) and 1 is the unit vector. In a three-dimensional system the square is replaced by a cube of the same side length and ξ and 1 have three instead of two components. There is a huge but finite number of neighboring states n, each of which are characterized by a distinct position inside the square $\mathbf{R}$ (see Figure 2). Since each of these N_R states may be reached from m with equal probability, α_{mn} is given by [11]

$$\alpha_{mn} = 1/N_R \qquad\qquad \mathbf{r}_i^m \in \mathbf{R}$$
$$\tag{22}$$
$$\alpha_{mn} = 0 \qquad\qquad \mathbf{r}_i^m \in \mathbf{R}$$

To construct the transition matrix π in eqs. (17) and (18) with the help of eq. (21), two cases have to be distinguished [11]:

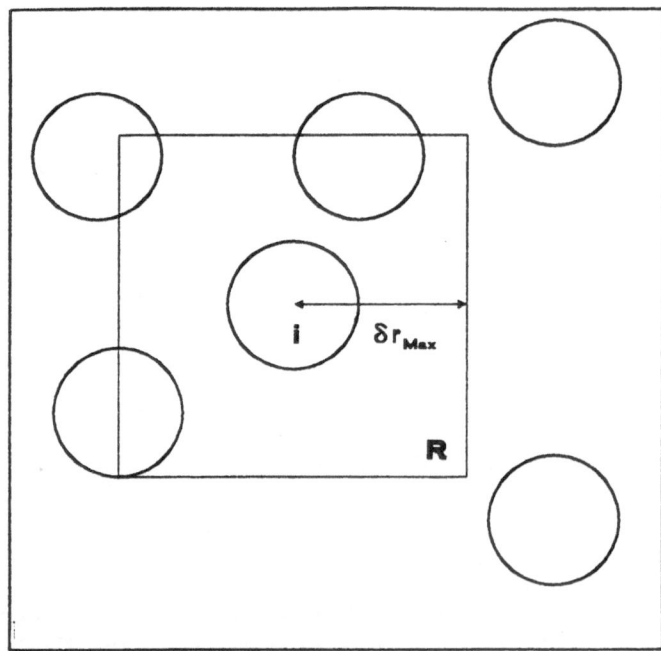

Figure 2. Generation of a Monte Carlo trial state in a two-dimensional system. Particle i is randomly displaced on a square of side length δr_{Max} (cf. Figure 4.3 in [11]).

$$\Delta U_{nm} = U_n - U_m \leq 0 ; \qquad \rho_n \geq \rho_m$$

$$\Delta U_{nm} \qquad\qquad \leq 0 ; \qquad \rho_n \geq \rho_m \tag{23}$$

where

$$\Delta U_{nm} = \sum_{\substack{j \neq 1}}^{N} [U(r_{ij}^n) - U(r_{ij}^m)] \tag{24}$$

denotes the change in potential energy between states m and n if we restrict ourselves to pairwise additive interaction potentials. State n is accepted with probability 1 as a new member of the Markov chain if $\Delta U_{nm} < 0$; this state will only be accepted with a smaller probability

$$\frac{\rho_n}{\rho_m} = \frac{Z_{NVT}^{-1} \exp[-\beta U_n]}{Z_{NVT}^{-1} \exp[-\beta U_m]} = \exp[-\beta U_{nm}] \tag{25}$$

if $\Delta U_{nm} > 0$ (see second of the eqs. (19)). Note, that this ratio does not depend on the configurational integral.

When it was first proposed in 1953 the validity of the Metropolis algorithm was not immediately accepted but became a controversial subject between Rosenbluth and Kirkwood who basically questioned the way in which Markov chains were generated numerically. Kirkwood's criticism concerned the "randomness" of the succession of configurations generated by the Metropolis algorithm but this matter could soon be settled. For more details about the early history of MC calculations we refer the interested reader to the excellent article by Wood [26].

So far only MC in the canonical ensemble has been considered. In this ensemble a phase space trajectory is generated by random displacements of particles. In other statistical mechanical ensembles generating a phase space trajectory requires more than just random displacements. For example, in the grand canonical ensemble particles are created and destroyed in addition to being randomly displaced. Here one has to make sure that both processes are equally weighted. Similar considerations apply to polyatomic molecular fluids where rotations of molecules have to be taken into account [11].

III.2. Computational Aspects

From eqs. (23) and (24) it is obvious that generating a realization of a Markov chain in the canonical ensemble requires computation of pair interactions between particles. Since we are investigating microscopic samples containing typically from several tens up to a few thousand particles per roughly 10^{-27} to 10^{-26} m^3, this calculation calls for additional considerations.

Suppose that we have a two-dimensional system and particles are put in a square of side length L (see Figure 3). In order to specify the thermodynamic state of this system in the canonical ensemble, we have to fix the density somehow. In principle

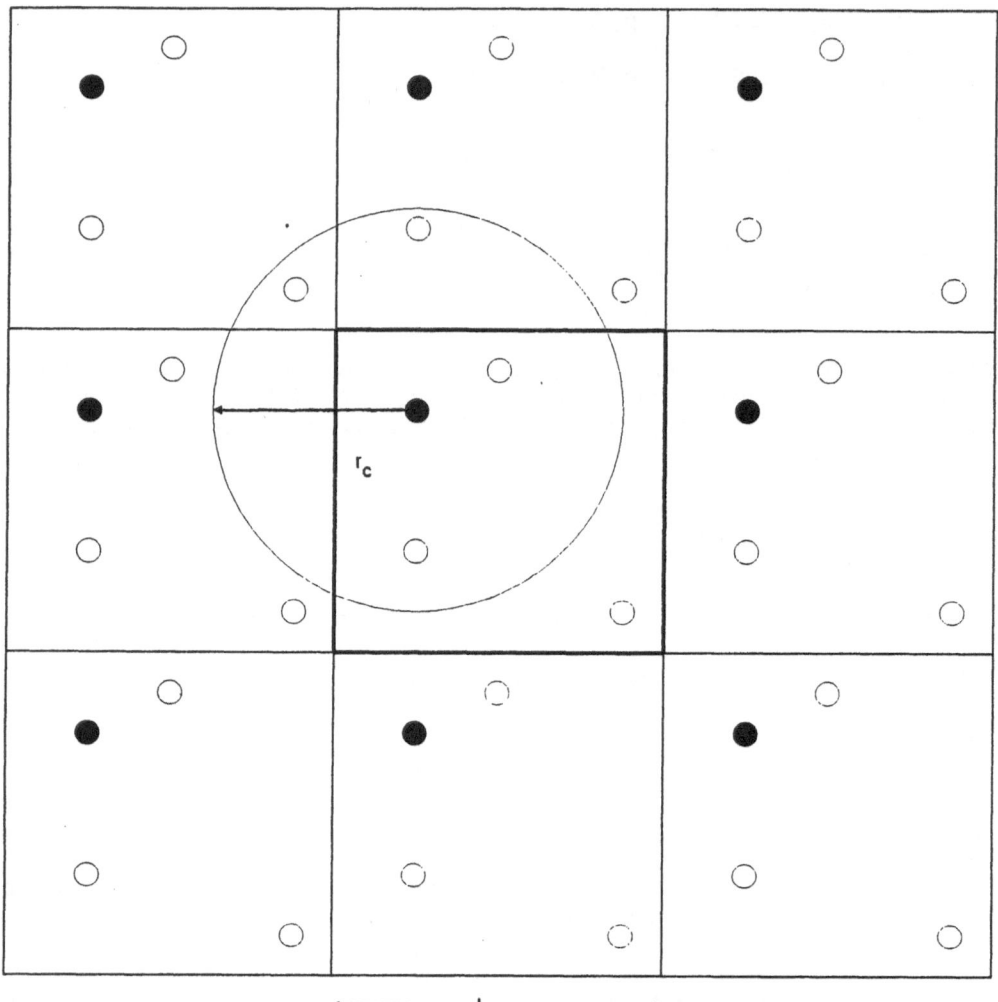

Figure 3. Periodic boundary conditions and minimum image convention in a two-dimensional system. r_c denotes the radius of a cutoff circle centered on a particle and L is the side length of the simulation cell.

this could be achieved by solid walls as indicated by the full lines surrounding the central square in Figure 3. Unfortunately, the surface to volume ratio would be rather unfavorable, that is even for very short-ranged potentials an intolerably large fraction of particles would interact with the walls. System properties would then be dominated by undesirable surface effects.

One possible solution of this problem is to introduce periodic boundary conditions. The central square is thus surrounded by replicas containing identically the

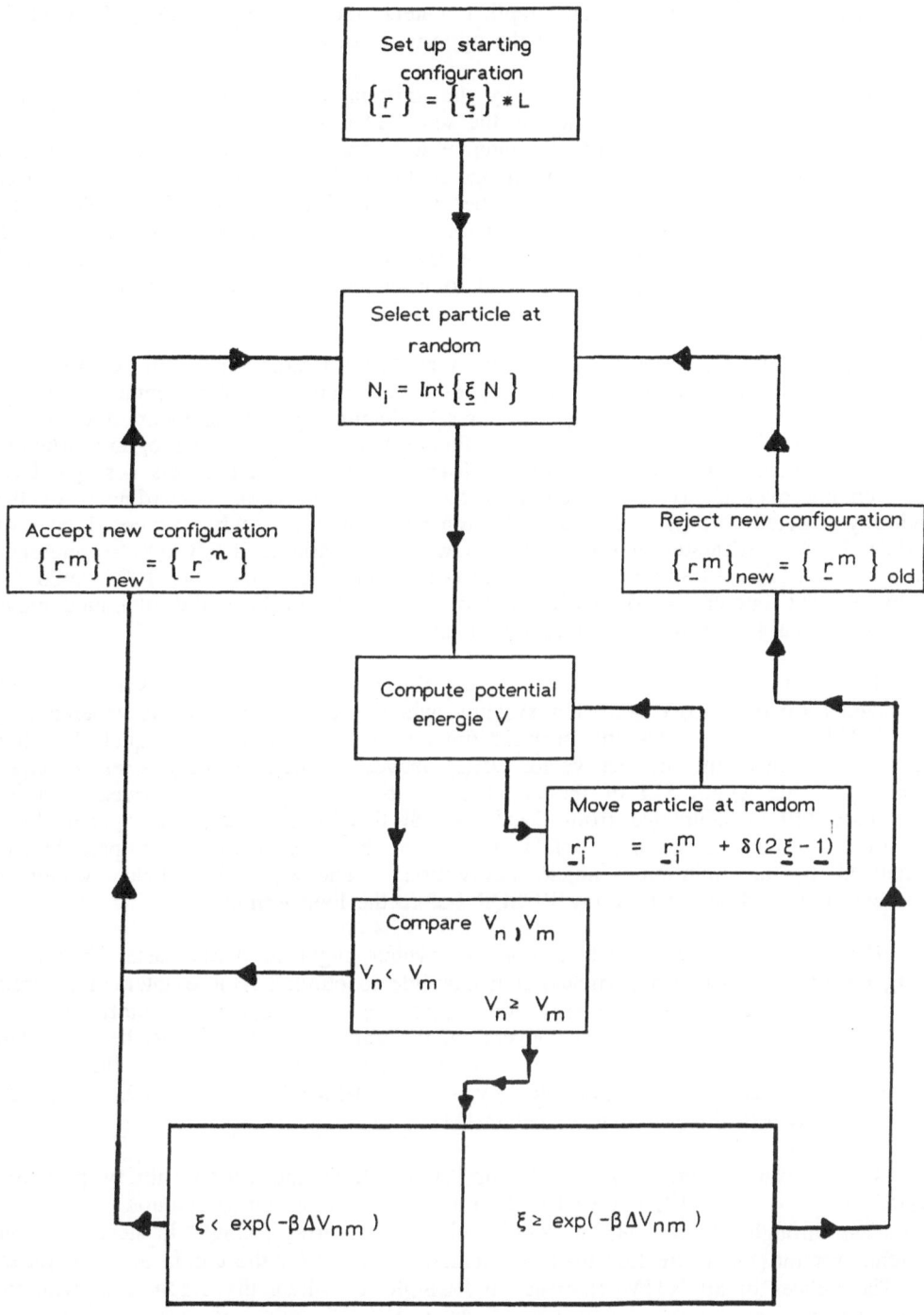

Figure 4. Flow diagram for a canonical ensemble Monte Carlo program; V denotes potential energy.

same configuration as the central square. Constant density is maintained because an image particle from the relevant replica enters the central square whenever a corresponding real particle leaves it in the opposite direction.

To avoid spurious effects caused by the artificial periodicity one also limits the range of interaction between particles by the so-called **minimum image convention**: A particle in the central square must not interact with one of its neighbors in the central square and an image of this neighbor in one of the replicas at the same time but only with the closer one , regardless of whether this closer one is a real particle or an image of a real one (see Figure 3). As before , the square is replaced by a cube in a three dimensional system. Other shapes for the simulation cell are also possible and have been used but special care is needed to implement periodic boundary conditions and minimum image convention [27-29].

In terms of CPU time the computation of (pair) interactions is the central task during any MC calculation (see flow diagram in Figure 4). Any optimization for vector or parallel processors has to focus predominantly on that segment of computer code in which this calculation is done. In Figure 5 an example of a program structure for the innermost part of the relevant subprogram is given. It contains a single DO-loop running over all (N - 1) particles according to eq. (24). Coordinates of the randomly selected particle i have been interchanged previously with the coordinates of particle 1 in a separate subroutine to allow index variable J always to reference contiguous storage locations. Otherwise vectorization of the loop 100 would be inhibited. The interchange of particle coordinates is undone once the attempted move of particle i has finally been accepted or rejected.

The next two lines of code represent the minimum image convention in an optimized form for CRAY computer systems, where vector inline code is generated by the FORTRAN compiler for the intrinsic function AINT. AINT(XX) equals 0. when the x component of the distance vector DX(J) between particles i and j is less or equal to half the side length SEI of the simulation cube and equals 1. otherwise. In the latter case SEI is subtracted from DX(J), which thereby becomes a component of a new distance vector between particle i and one of the periodic images of particle j as demanded by the minimum image convention. The squared distance vector is computed next and stored on array RCSQR before the loop terminates.

For a reasonably short-ranged potential which ought to decay faster than r^{-3} (r being the distance between particles) it is possible to eliminate those interactions from the computation of potential energy for which r exceeds a certain cutoff value r_c. According to the minimum image convention, the upper limit on r_c must be L/2. The precise value of r_c depends on a variety of criteria. For an atomic fluid whose interparticle interactions are represented by a Lennard-Jones(12,6) potential, r_c typically ranges from two to three atomic diameters [11].

In the example program (see Figure 5) a call to the CRAY library programs WHENFLE and GATHER allows for the elimination of distant interactions. The first call scans through the (N - 1), i.e. NTEM, elements of array RCSQR in steps of 1 and searches for those that are less than or equal to r_c^2, i.e RCQ, the cutoff distance square d. The indices of all NTIM elements in RCSQR for which the comparison with the scalar RCQ is successful are stored on array INDEX, which is used to compress RCSQR onto itself during the next call to GATHER. The potential energy is then computed with this reduced number of elements in a separate subroutine.

There are various other ways of eliminating particle interactions from the computation of potential energy. These techniques have mainly been developed for

```
                        .
                        .
                        .

        SEQ      = 2. /SEI
        NTEM     =      NTEI - 1
        DO 100 J = 2,NTEI
        DX      (J) = X (1) - X(J)
                        .
                        .
            Y, Z components
                        .
                        .
        XX          =DX(J) * SEQ
                        .
                        .
            Y, Z components
                        .
                        .
        DX      (J) =DX(J) - SEI *AINT(XX)
                        .
                        .
            Y, Z components
                        .
                        .
        RCSQR(J) =DX(J)**2+DY(J)**2+DZ(J)**2
  100   CONTINUE
        CALL  WHENFLE(NTEM,RCSQR,1,RCQ,INDEX,NTIM)
        CALL  GATHER(NTIM,RCSQR,RCSQR,INDEX)
                        .
                        .
  Compute potential energy with NTIM elements of squared distances
                     stored on RCSQR
                        .
                        .
        RETURN
        END
```

Figure 5. Typical vectorizable program structure for the innermost loop of a Monte Carlo program. The present example is designed for CRAY computers.

MD programs rather than for MC but may also be used in revised form for the latter. Most of them are based on a kind of sorting procedure [30-32].

The above program structure can also be rewritten for a CYBER 205 or ETA 10 system, where the special routines WHENFLE and GATHER would have to be replaced by BIT vector oriented, so-called Q8-routines like, for instance, Q8VCMPRS, which is the analogue of GATHER on a CRAY [33].

Table 1. CPU-time for 10^3 configurations on a T800 transputer network (from [34]).

N	205/vec. sec	205/seq. sec	1 x T8 sec	4 x T8 sec	8 x T8 sec	16 x T8 sec
108	2,25	29,6	134,7	37,8	21,2	13,3
256	3,69	68,9	318,4	85,9	46,0	25,2
500	6,04	132,9	620,4	164,9	84,8	46,0
864	9,59	228,6	1069,0	282,3	144,6	75,4
1000	10,09	264,2	1236,0	326,8	167,3	88,9
2048	21,2	536,5	2535,0	667,3	337,6	172,3

While most computer simulations are still performed on vector machines or other big scalar mainframes, recent developments in parallel computing and transputer technology make parallelization of statistical mechanical computer programs an attractive possibility. The structure of such a program may differ substantially from one highly adapted to a particular vector machine. Parallelizable modifications beyond vectorizability may also be desirable if a multiprocessor machine like the CRAY-Y/MP can be employed. For a conventional transputer system, however, the program structure outlined above appears well suited for parallelization (after specific CRAY features are replaced), because the entire innermost loop can be split into smaller independent "subloops". To each subloop a transputer may be assigned to compute certain elements of the squared distance vector. Care must be taken to ensure an effective communication between individual transputers. For example, the number of arithmetic operations (i.e. the number of particle interactions) to which a single transputer is dedicated should not be too small [34], since computational speed increases linearly with the number of transputers employed only up to a certain limit. Beyond that point chaining of the results sent back from each transputer may take too long compared with the calculation within a transputer. Table 1 illustrates the speed up realized from differently sized transputer systems compared with a canonical MC program fully vectorized for a CYBER 205 [34]. A comparison of CPU times reveals that a system of sixteen T800 transputers achieves a computational speed that makes MC calculations feasible even for large systems containing as many as 2048 particles.

Although parallelization of the potential energy calculation discussed above is probably the most efficient, one can envision alternative approaches. One possibility is to generate as many Markov chains as there are transputers in the network. This would increase statistical accuracy for certain system properties with a substantially reduced length of each individual Markov chain. There are, however, situations in which such a reduction is inappropriate. These are frequently met in conjunction with transitions, where it is sometimes difficult and costly to ensure ergodicity of a single Markov chain in a MC run (see Sec. III. 3). This is plausible because different phases may be represented as more or less isolated regions in phase space. The system undergoes a phase transition only along a very narrow path connecting these regions, which is difficult to find by a trial and error procedure, even if importance sampling is employed. Reducing the length of a Markov chain would probably not allow the system to pass such a bottleneck and thus the phase transition would not occur (see also Figure 2.1 in [11]).

Another possible alternative of parallelization could be based upon the movement of several particles at once. In principle, all particles in the system may be displaced randomly in a single step but the transition probability between states m and n will likely be too small. For a short-ranged potential, however, it should be possible to move a certain fraction of particles at once, subject to the constraint that no two particles in this group interact with each other. Thus, the random displacement of a particle in one region and its associated change in potential energy does not affect the change in potential energy due to the displacement of a particle in another region and vice versa. In effect, the entire system is split into smaller subregions or subcells for which neighboring substates n can be generated independently (i.e. in parallel). Each subregion has a separate transition probability and the random movements of the entire group of particles are accepted or rejected for the individual substates according to the Metropolis algorithm outlined above.

IV. MONTE CARLO CALCULATIONS IN THE GRAND CANONICAL ENSEMBLE

In the previous section of this article aspects of the most fundamental version of MC in the canonical ensemble for atomic fluids were discussed. Historically this was more or less the way in which pioneers in this field, like Metropolis et al. [25] or Wood et al. [13], did their early work in the late 1950's. Since then a lot of work has been invested to extend the MC method in various directions, so that nowadays rather complex systems and difficult physical problems can be tackled. This was made possible not only by the enormous, almost astronomical increase in computer power over the years but also by the great inherent conceptual flexibility of the method itself.

In principle it is possible to design MC algorithms which allow for realizations of Markov chains in accordance with almost any statistical mechanical ensemble. Besides the canonical the most frequently used other ensembles are the isothermal isobaric (N,P,T) and the grand canonical one in which a thermodynamic state is specified by fixed values for chemical potential, volume and temperature [7]. Depending on the physical problem of interest, one ensemble may be more convenient than another. The formal equivalence of all statistical mechanical ensembles guarantees that results obtained in one ensemble will be the same in any other, except for small correction terms of order N^{-1} [35] due to the finite system sizes studied in practice.

From the standpoint of computational convenience, however, a careful choice of the most appropriate ensemble can make an important difference. For example, in the grand canonical ensemble it is possible to calculate directly thermodynamic quantities like the constant volume heat capacity C_V or the isothermal compressibility κ from fluctuations of particle number and energy, since N is no longer fixed during the MC run [7] (see Sec. IV. 3). Grand Canonical Ensemble MC (GCEMC) is also advantageous for the study of phase equilibria or phase transitions: For a given chemical potential μ the system attains the thermodynamically most stable phase provided the Markov chain is ergodic. In the canonical ensemble, on the other hand, μ must be computed [36] for a variety of different thermodynamic states before we can decide which one is most stable at a particular μ. Thus, the computational effort is substantially larger in the canonical ensemble.

In the remainder of this article we intend to review the GCEMC method. We will also demonstrate its particular usefulness for a study of liquid-gas phase transitions in an anisotropic fluid, where the requirement of computer time can be huge.

IV.1. The Model System

A fluid in the presence of an external field is generally anisotropic. The anisotropy is primarily reflected, for example, in the local density $\rho^{(1)}(\mathbf{r})$ which is non-uniform. The field causing such anisotropy may be due to imposed electric or magnetic fields or to particles belonging to a distinctly separate phase in contact with the fluid phase of interest. If we **define** our system to include only particles in the fluid phase, then these particles may be viewed as subject to an external field due to particles in that distinctly separate phase. This is the situation in porous media, where a fluid is confined by solid phases (e.g. walls). The different physical behavior of particles belonging to either phase suggests that we treat fluid particles as though they are acted upon by an external, time independent field. This can be justified because typical thermal fluctuations in a fluid occur on a much longer time scale than in a solid.

Two different geometries of model pores have been studied theoretically, namely cylindrical pores and slit pores. In a slit pore the fluid is confined between two walls which are taken parallel to each other and perpendicular to the z axis of the "laboratory" coordinate frame (see Figure 6). To represent layers of infinite extent, we impose periodic boundary conditions in the x and y directions. The space between the solid layers is occupied by N particles. This geometry resembles a real experimental situation in clay-water systems, where certain clays can form large sheets which are flat on an atomic scale [37]. Compared with real clay-water systems the ones usually studied theoretically are rather idealized versions of the former because fluid particles are mostly assumed to be nonpolar and the interparticle interaction is modelled by spherically symmetrical, pairwise additive potential functions of the Lennard-Jones(12,6) type. Since this potential function is short-ranged, interactions are cut off if the

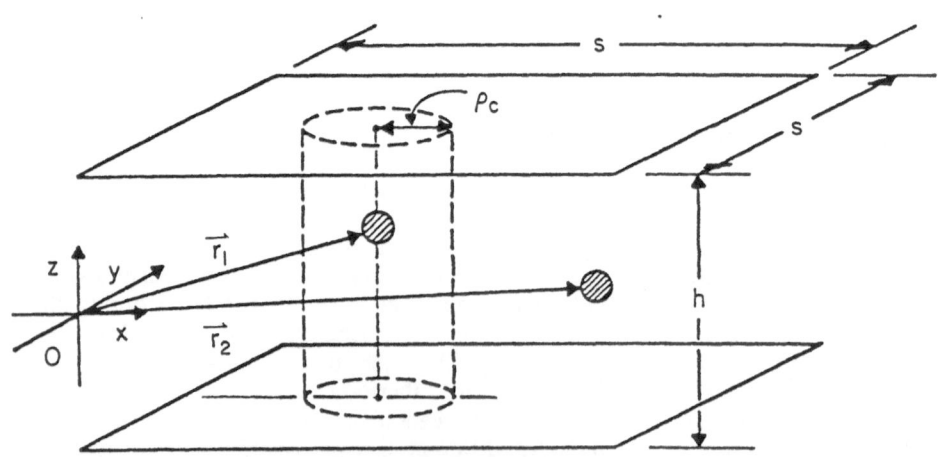

Figure 6. Schematic of the model slit-pore, showing planes of solid layers, cutoff cylinder and coordinate system employed in evaluating corrections to the configurational energy; ρ_c is the cutoff radius.

distance between two particles located at $\mathbf{r}_1$ and $\mathbf{r}_2$ (see Figure 6) is larger than the radius ρ_c of a cutoff cylinder of infinite height centered on particle 1.

In the experiments the pore fluid is in thermodynamic equilibrium with the bulk fluid. The well known condition for such an equilibrium is equality of chemical potentials μ' and μ'' for the two phases ' and ". Therefore, it is most convenient theoretically to fix the chemical potential for the pore fluid, μ', in a GCEMC calculation in order to compare the results with the ones obtained for the corresponding bulk fluid at the same chemical potential $\mu'' = \mu'$. Alternatively one could keep the number density fixed at a certain value for bulk and pore fluid and try to estimate μ' and μ'', for example, by means of the test particle method in a canonical MC (or microcanonical MD) study [38]. However, several such runs would be needed to find those number densities for which the calculated chemical potentials equal each other. In the grand canonical ensemble only two runs are necessary since a fixed value for m is input information for the bulk as well as for the pore system and the corresponding number densities are calculated from each run accordingly. Thus, in terms of computer time the required efforts are much smaller for a GCEMC than for a MC study of the same system in the canonical ensemble.

At this point it should be noted that the fluid in our model slit pore is not in direct contact with any bulk reservoir. Edge effects resulting possibly from interactions between pore fluid particles with reservoir particles are not accounted for. In spite of all these simplifications it is nevertheless found that the model system behaves realistically enough to reveal properties in qualitative agreement with experimental results for much more complex systems. For example, the so-called solvation force f_s, which is the force exerted by fluid particles on the solid wall turns out to be an oscillatory function of h (i.e. the distance between the walls) both experimentally [39] and theoretically [40].

Snook and van Megen have shown how this oscillatory behavior relates to the arrangement of fluid particles in layers parallel with the walls [41]. The number of layers increases with h; at certain critical values of h nearly a complete new layer of fluid "pops in", leading to the experimentally observed oscillations in $f_s(h)$ (see also Figure 5 in [42]).

Figure 7 displays plots of $\rho^{(1)}(z)$ as a function of position between walls for various wall separations. While a strongly anisotropic fluid may be observed for narrow pores, the layering remains only for portions of fluid closest to either wall as the pore becomes very wide. These observations remain qualitatively the same regardless of the details in the fluid-solid interaction potential.

For that reason structureless (i.e. smooth) walls are commonly used. Here the fluid-solid interaction potential depends only on the relative z coordinate of fluid particles with respect to the walls, which makes such a potential computationally convenient. The most frequently used potential function of this type was suggested by Steele to mimic argon-graphite adsorption [43]:

$$U(z) = 2\pi\varepsilon \left(\frac{4}{5}\right) \left[\frac{2}{5}\left(\frac{\sigma}{z}\right)^{10} - \left(\frac{\sigma}{z}\right)^{4} - \frac{\sqrt{2}}{3\left(\frac{z}{\sigma} + 0.61\sqrt{2}\frac{z}{\sigma}\right)^{3}}\right] \qquad (25)$$

Various modifications of this potential have also been employed [44,45]. In 1981, Toxvaerd concluded that replacing smooth walls by structured ones with individual

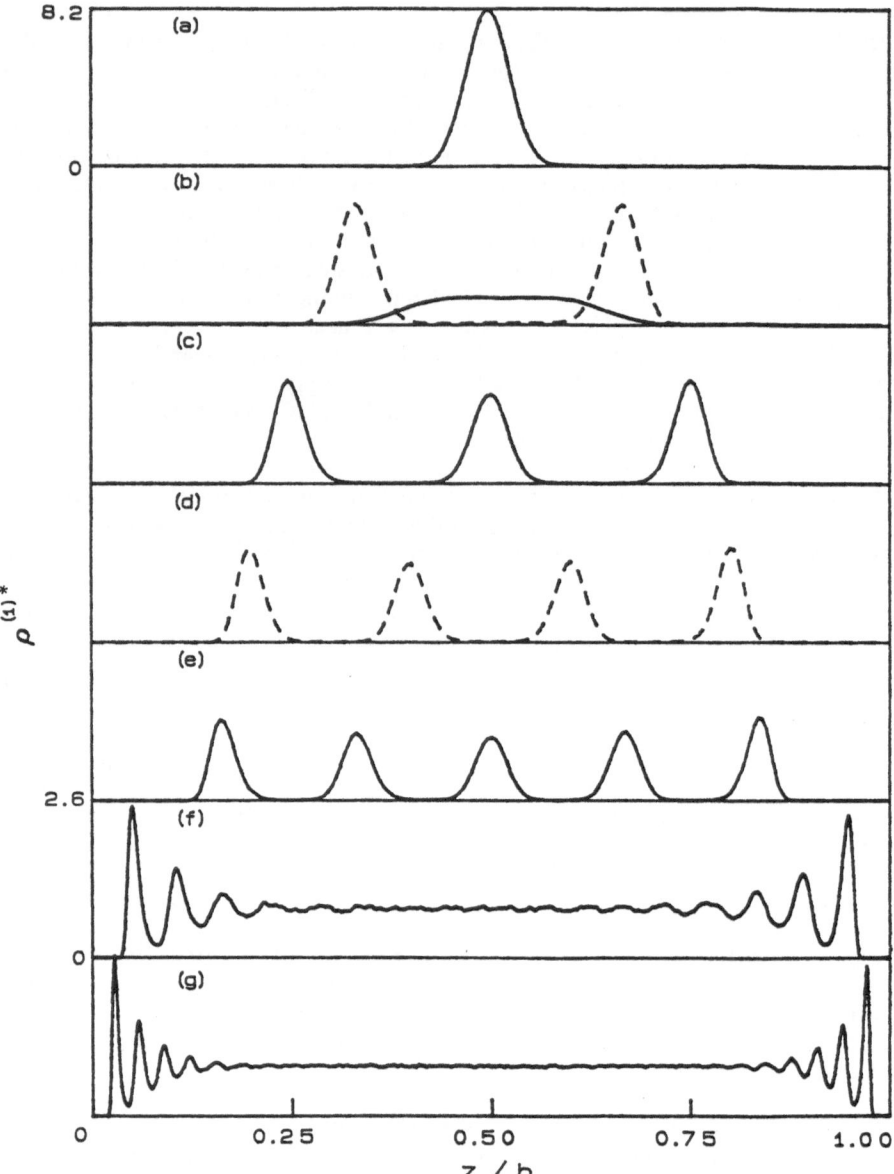

Figure 7. Local density profiles as a function of inter-wall separation h* for pore fluid at a temperature T* of 1.0. The surface density of wall atoms is $N_s/s^{*2} = 0.7827$, where $N_s = 50$, the number of wall atoms. Panel a, h* = 1.37, $\mu^* = -9.00$; b, h* = 2.20, $\mu^* = -9.26$; c, h* = 3.05, $\mu^* = -10.29$; d, h* = 3.90, $\mu^* = -9.83$; e, h* = 4.90, $\mu^* = -10.00$; f, h* = 16.5, $\mu^* = -10.29$; g, h* = 30.0, $\mu^* = -10.29$. In panels a-e the scale of $\rho^{(1)*}$ ranges from 0.0 to 8.2; the solid and dashed curves respectively correspond to walls in and out of registry (see eq. (27)). In panels f and g the scale of

interaction sites in the walls does not significantly alter system properties [46]. Later work showed that this conclusion was premature [47,48]. There it was found that it is possible for a fluid in a pore with structured walls to "freeze" (even though this pore "fluid" is in thermodynamic equilibrium with bulk fluid) whereas freezing does not occur in the smooth wall pore. The solid-like structure of the system is reflected in the very pronounced layering (see Figure 7) of the fluid parallel with the walls. The freezing occurs epitaxially, the walls acting as templates for the structure of the pore solid (see also Figure 3 in [47]). Vestiges of the wall structure remain visible in the

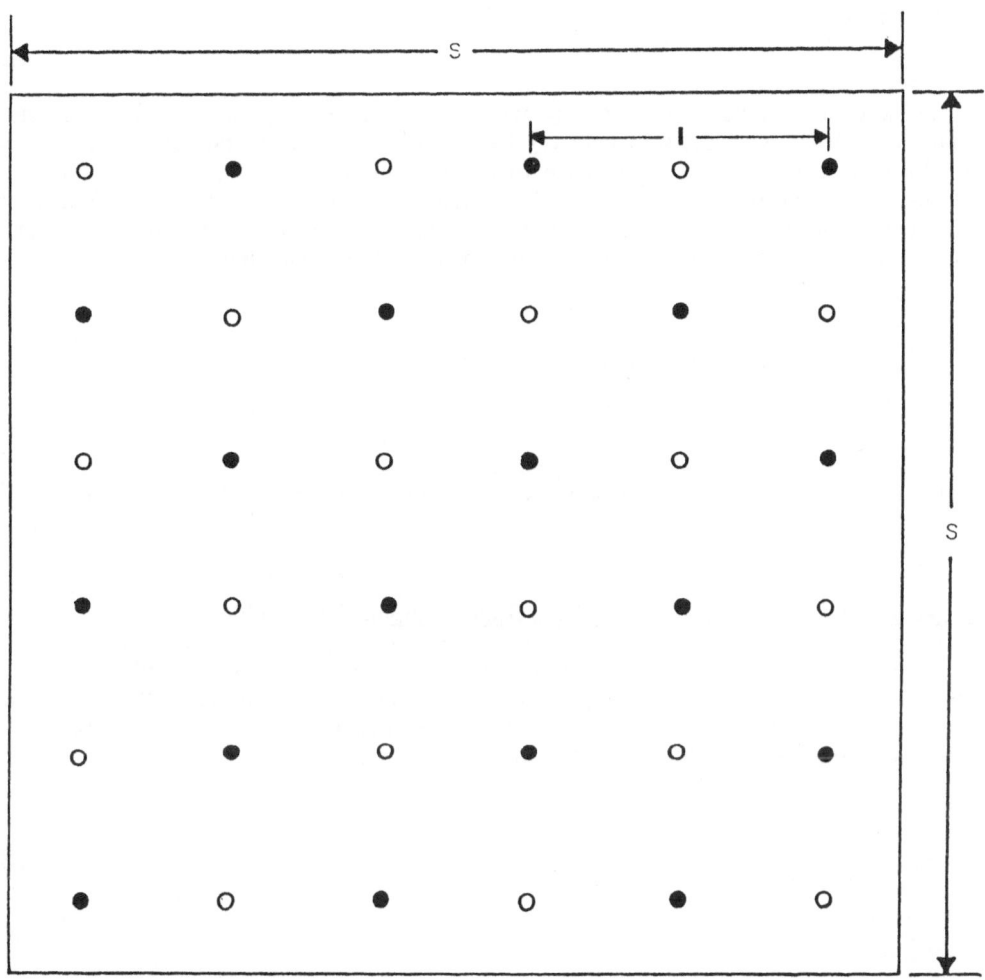

Figure 8. A square unit cell of the fcc (100) plane of the fcc lattice. The filled and open circles respectively represent atoms in the solid layers at z = 0 and z = h.

lateral structure of the portion of fluid closest to the walls up to very large wall separations h. Similar observations have been made in light scattering experiments on fluids next to solid substrates composed of the same molecular species [49].

The results to be presented in this article have also been obtained with structured walls. In particular, the (100) plane of a fcc lattice (Figure 8) is employed here. Only one layer of solid atoms is put on top and on bottom of the slab of fluid (see Figure 6). Positions of wall particles in the upper (z = h) wall are related to those in the lower (z = 0) one according to

$$x_i^h = x_i^0 + \alpha l$$

$$y_i^h = y_i^0 \qquad ; \quad i = 1, \ldots\ldots, N_s \quad \text{(number of solid particles)} \qquad (26)$$

$$z_i^h = z_i^0 + h$$

α is the so-called registry parameter by which the relative alignment of wall particles can be varied from completely **in registry** (α = 0.0) to completely **out of registry** (α = 0.5). Its precise value is of importance to the formation of pore solids [45,47,48] but is of no great relevance to liquid-gas phase transitions in porous media. The results we present below have been obtained with a structured wall where the fluid-solid and fluid-fluid interactions are modelled by Lennard-Jones(12,6) potential functions

$$U_{FS}(r_{ij}) = 4\varepsilon \left[\left| \left(\frac{\sigma}{z} \right) \right|^{12} - \left| \left(\frac{\sigma}{z} \right) \right|^{6} \right] \qquad (27)$$

with identically the same potential parameters ε and σ (ε/k_B = 119.8K, σ = 3.4 × 10^{-10} m).

IV.2. A Monte Carlo Algorithm For The Grand Canonical Ensemble

In this section an algorithm is described which generates a realization of a Markov chain compatible with the grand canonical ensemble. This algorithm is similar to the one already discussed for the canonical ensemble, because the partition functions in the two ensembles are closely related via

$$\Xi_{\mu VT} = \sum_N e^{\beta\mu N} Q_{NVT} \qquad (28)$$

The summation over N indicates that the number of particles is not a fixed thermodynamic state variable. An average quantity $\langle F_{\mu VT} \rangle$ can be obtained in principle from an expression similar to the one given in eq. (12), namely

$$\langle F \rangle_{NVT} = \frac{\displaystyle\sum_{N=0}^{\infty} V^N \frac{\exp[\beta\mu N]}{N!\Lambda^{3N}} \frac{1}{\tau_N} \sum_{i=1}^{\tau_N} F(r_i^N) \exp[-\beta U(r_i^N)]}{\displaystyle\sum_{N=0}^{\infty} V^N \frac{\exp[\beta\mu N]}{N!\Lambda^{3N}} \frac{1}{\tau_N} \sum_{i=1}^{\tau_N} \exp[-\beta U(r_i^N)]} \tag{29}$$

As before, importance sampling has to be employed to evaluate $\langle F_{\mu VT} \rangle$.

Several ways of generating a realization of a Markov chain in the grand canonical ensemble have been proposed in the literature [50,51]. The one we adopt here follows essentially suggestions by Norman and Filipov [52], who split the generation of a new configuration in the Markov chain into three different steps:

i) particle displacement at random
ii) creation of a particle at a randomly chosen point r_i^n
iii) destruction of a randomly chosen particle

where ii) and iii) are attempted with equal probability.

Following Adams [53], we also introduce a new quantity

$$B = \beta\mu - \ln(\Lambda^3/V) = \beta\mu' + \ln\langle N \rangle \tag{30}$$

where

$$\Lambda = \hbar (2\pi\beta/m)^{1/2} \tag{31}$$

again denotes the thermal de Broglie wavelength and μ' is the excess chemical potential. With this definition of B it can be shown that creation and destruction of a particle are accepted with probabilities

$$P_{\pm} = \begin{cases} 1 & ; \ r_{\pm} \geq 1 \\ r_{\pm} & ; \ r_{\pm} < 1 \end{cases} \tag{32}$$

where + refers to creation and - to destruction of a particle. r is defined as

$$r_+ = \exp[-B + \ln N_m + \beta\Delta U_{nm}] \tag{33}$$

and

$$r_- = \exp[-B - \ln N_m - \beta\Delta U_{nm}] \tag{34}$$

where m and n refer to the old state and the new trial state. As in the canonical MC method (see Sec. III. 1) random displacements are accepted with a probability

$$P_\pm = \begin{cases} 1 & ; \Delta U_{nm} < 0 \\ \exp[-\text{ß}\Delta U_{nm}]; & \Delta U_{nm} \geq 1 \end{cases} \tag{35}$$

By complete analogy with eq. (23) ΔU_{nm} denotes the change in potential energy between configurations m and n due to fluid-fluid and fluid-solid interactions:

$$\Delta U_{nm} = U_n - U_m + \Delta U_c \tag{36}$$

The third term in this equation, ΔU_c, is a correction because of the change in density by $\pm 1/V$ between configurations m and n when a creation-destruction attempt is made. ΔU_c is calculated as a sum of two contributions, namely

$$\Delta U_{c,FF} = \frac{1}{2} \int_V dr_1 \int_{V'} dr_2 \; U_{FF}(r_{12}) \; [\rho_n^{(2)}(r_1,r_2) - \rho_m^{(2)}(r_1,r_2)] \tag{37}$$

$$\Delta U_{c,FS} = \frac{1}{2} N_S \int_{V'} dr_2 \; U_{FS}(r_{12}) \; [\rho_n^{(2)}(r_1,r_2) - \rho_m^{(2)}(r_1,r_2)] \tag{38}$$

Note, that in eq. (37) the cutoff cylinder is centered on a solid atom as the reference. By symmetry, every solid atom contributes equally to the correction. The integration on r_1 is over the region V containing the system and that on r_2 extends over the infinite region V' exterior to the cutoff cylinder (see Figure 6). The integrations are so restricted that the tip of $r_{12} = r_1 - r_2$ lies outside the cutoff cylinder. Under certain assumptions the correction terms can be evaluated analytically (see appendix in [45]). From the eqs. (35) and (36) it is also evident that both terms vanish for the random displacement of a particle (i.e. the canonical step i)) because then $\rho_m^{(2)} = \rho_n^{(2)}$.

The creation-destruction steps in GCEMC cause an additional problem not present in canonical MC. While it is possible to adjust the rate of acceptance of trial configurations in the generation of the canonical Markov chain by varying δr_{Max}, the side length of the displacement cube centered on the randomly selected particle i, no such adjustment is possible for the creation-destruction steps in the GCEMC procedure. In the latter case the acceptance rate is determined exclusively by the physical conditions, i.e. the thermodynamic state of the system. If the system is rather dense, for instance, it can be very difficult to create or destroy particles. Many unsuccessful attempts may be necessary before one is finally accepted. At typical liquid densities acceptance rates for creation and destruction of particles may be as small as 0.1% whereas it is typically 50% for the canonical steps. These small acceptance rates necessitate the generation of several millions of configurations to ensure ergodicity of the Markov chain.

IV.3. An Illustration: Liquid-gas Phase Transitions in Slit Pores

The quantity of primary interest in the following discussion of the liquid-gas phase transition is the average number density

$$n = \langle N \rangle / s^2 h \tag{39}$$

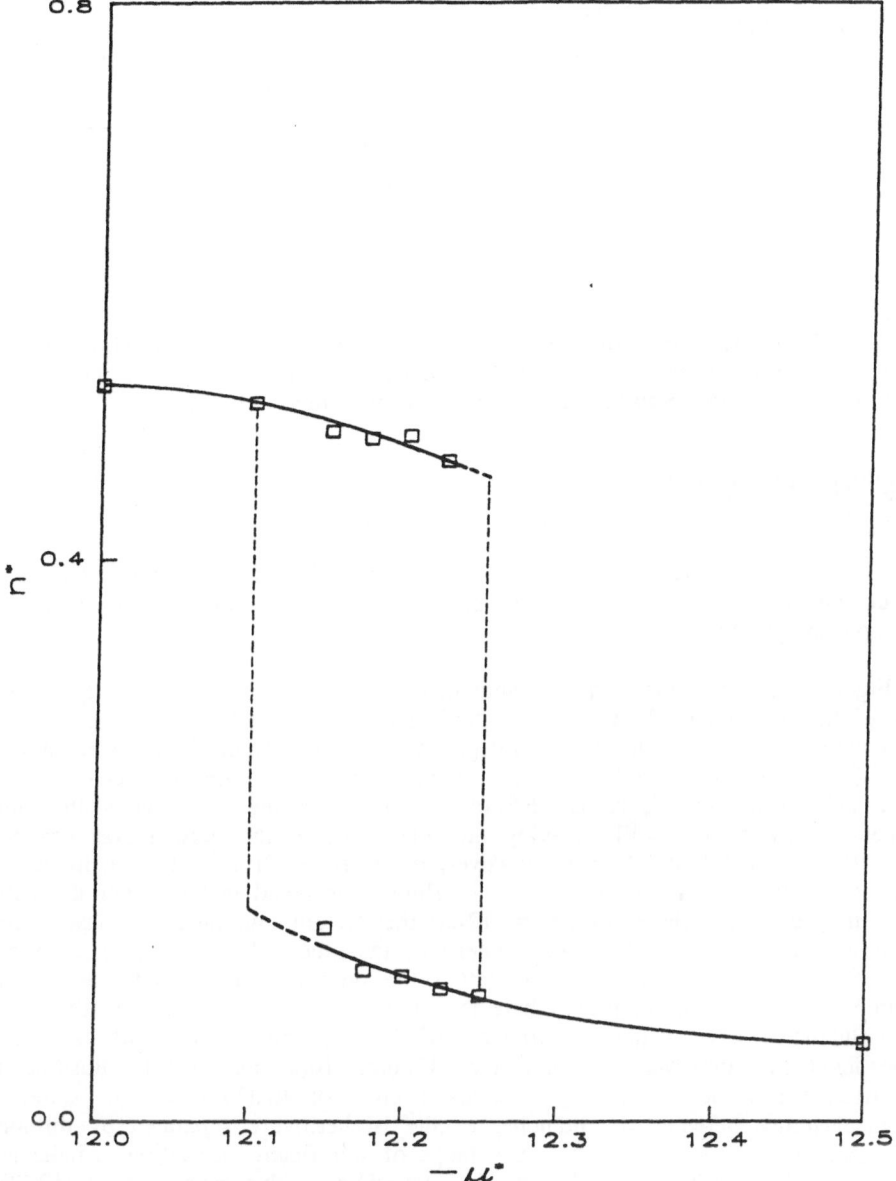

Figure 9. Sorption isotherms for pore at h* = 4.90, T* = 1.0 and s* = 7.9925. Vertical dashed lines indicate range of chemical potential within which hysteresis is observed.

Table 2. Reduced variables. $\varepsilon/k_B = 119.8$ K and $\sigma = 3\cdot405 \times 10^{-10}$ m are the Lennard-Jones potential parameters; k_B is Boltzmann's constant.

$h^* = h/\sigma$	$\mu^* = \mu/\varepsilon$
$s^* = s/\sigma$	$T^* = k_B\,T/\varepsilon$
$n^* = <N>\,\sigma^3/(s^2 h)$	$\rho^{(1)*} = \rho^{(1)*}\sigma^3$

where $<N>$ is the average number of particles computed in a GCEMC simulation for a particular set of parameters μ, V and T. To describe the structure of the pore fluid quantitatively, we also compute the local density from the expression

$$\rho^{(1)}(z) = <N(z)> / s^2\,\Delta\,z \qquad (40)$$

where $<N(z)>$ is the average number of particles within an imaginary slice of pore fluid centered on z and having width $\Delta\,z^* = 0.01$ (see Table 2 for definition of dimensionless quantities.)

Figure 9 displays the sorption isotherms (i.e. average number density, n, versus negative chemical potential, $-\mu^*$). Two distinct branches can be discerned. The upper branch corresponds to the lower density gas regime. For more details see Table 2 in [54]. That these states do indeed correspond to gas and liquid is borne out by the inplane pair correlation functions, $g^{(2)}$ (see [47]). On the gas branch the fluid is adsorbed mainly on the walls, leaving the inner part of the pore almost empty (see Figure 10). On the liquid branch, however, $\rho^{(1)}$ is more structured, with higher peaks corresponding to much higher density. The fluid is arranged in layers parallel with the walls. In the neighborhood of $\mu^* = -12.20$ the system undergoes a phase transition and 'jumps' from one branch of the isotherm to the other. For the particular pore size $s^* = 7.9925$, however, we cannot ascertain the precise value of μ^* at which the transition occurs. On the contrary, there is a non-zero range of values of μ^* where the system fluctuates between the two branches of the isotherm. Plots of n^* as a function of the length of the Markov chain (see Figures 10(a) and 11(a)) illustrate these fluctuations for values of μ^* near opposite ends of that range. In some cases subsequent jumps between the branches of the isotherm are separated by millions of Monte Carlo steps (see Figure 13). The range of significant fluctuation is indicated by dashed vertical lines in Figure 1. At the endpoints of this range, $\mu^* = -12.25$, the system jumps immediately (i.e. within a few hundred thousand Monte Carlo steps) to the liquid and gas branches, respectively, and remains there for at least another four to five million steps. For each value of μ^* in this range we obtain two values of n^*. The two different values of n^* plotted in Figure 9 were obtained by averaging separately over configurations belonging to the corresponding branches of the isotherm. We take the existence of this 'range of fluctuation' in the vicinity of the phase transition to be a manifestation of hysteresis within the context of the GCEMC method.

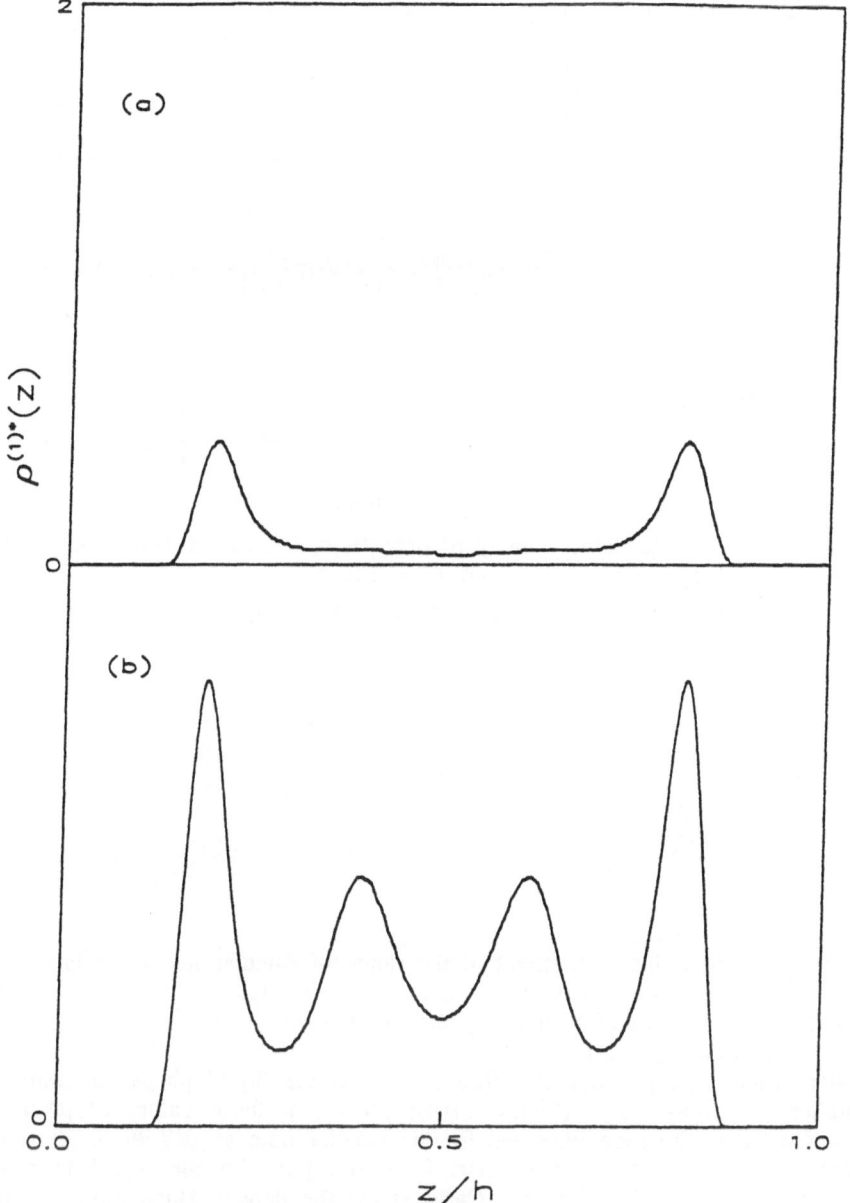

Figure 10. Local density $\rho^{(1)}(z)$ as a function of z/h at h* = 4.90, T* = 1.0 and s* = 9.591.   (a) μ* = -12.25.   (b) μ* = -12.175.

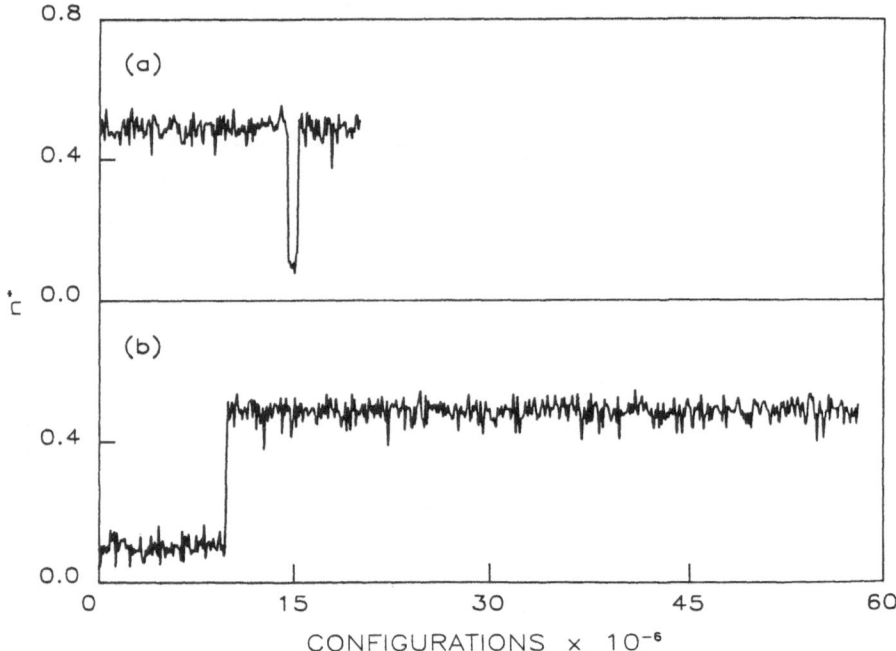

Figure 11. Average liquid density of pore fluid, n*, as a function of number of GCEMC steps at h* = 4.90 and T* = 1.0.

(a) s* = 7.9925, μ* = −12.175. (b) s* = 9.591, μ* = -12.175.

To attempt to reduce the extent of the range of fluctuations, we enlarged the pore to s* = 9.591. Figure 12(b) shows that at μ* = -12.225, starting from a liquid configuration, the system immediately jumps to the gas, where it stays for at least 22 million steps. Starting from a gas configuration at μ* = -12.175, on the other hand, the system finally jumps after 10 million steps to the liquid phase and remains there for another 48 million steps (Figure 11(b)). Both of these values of μ* are in the range where fluctuations are observed for the smaller pore s* = 7.9925. It should also be noted from Figure 12(a) that, apart from the jump to the liquid branch of the isotherm at about 18 million Monte Carlo steps, the density fluctuations in the vapor phase are significantly (and artificially) enhanced because of the smaller surface area of the pore walls. The run at μ* = -12.175 (Figure 11(b)) illustrates the enormous number of Monte Carlo steps that may be required to obtain the correct density as the phase transition is approached.

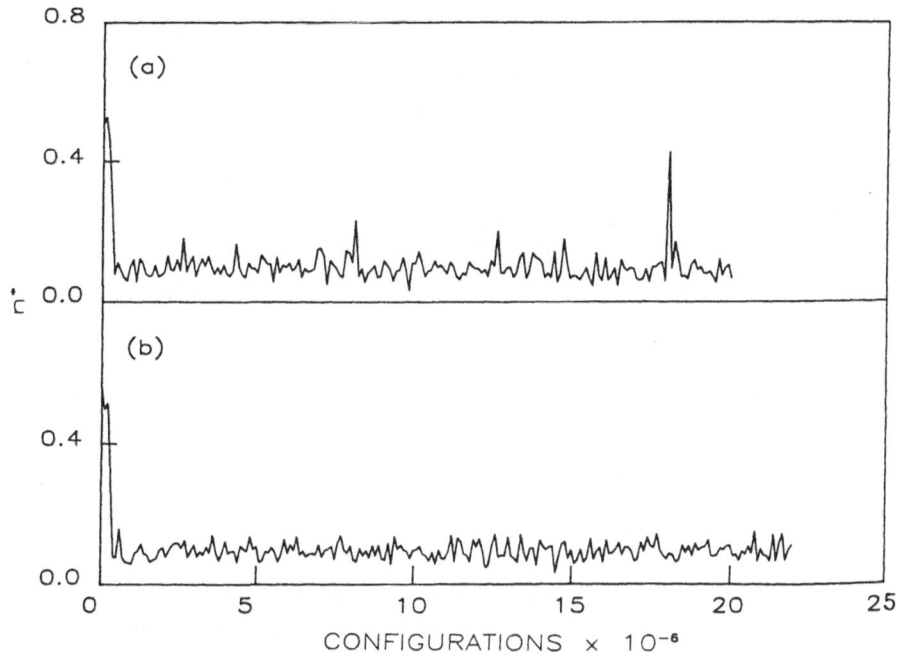

Figure 12. Same as Figure 11.

(a) $s^* = 7.9925$, $\mu^* = -12.225$.   (b) $s^* = 9.591$, $\mu^* = -12.225$.

None of the results thus far presented depends on the actual starting configuration. For values of $\mu^* = -12.20$ several GCEMC runs were carried out for $s^* = 9.591$. Starting from a liquid configuration (generated at $\mu^* = -12.15$), the system remains a liquid for at least 14 million steps. On the other hand, if it is started from a gas configuration (generated at $\mu^* = -12.25$) it remains a gas for at least 14 million steps. Apparently we must use even larger systems and longer runs to locate more precisely the transition.

Finally, in Figure 14 we plot the sorption isotherm for $h^* = 4.9$, using only the results for $s^* = 9.591$ in the immediate vicinity of the phase transition. We also plot the isotherms for $h^* = 3.05$ and $h^* = 2.15$. The isotherm for $h^* = 4.9$ appears to have a discontinuity at $\mu^* = -12.20$ approximately, whereas the other two isotherms are continuous. Between $\mu^* = -11$ and $\mu^* = -16$ we have found no range where either of the other two pores exhibits fluctuations in n^*. By means of the GCEMC method we have determined sorption isotherms for the rare-gas slit-pore model at the fixed temperature $T^* = 1.00$ and at several pore widths h^*. With increasing h^* the isotherm

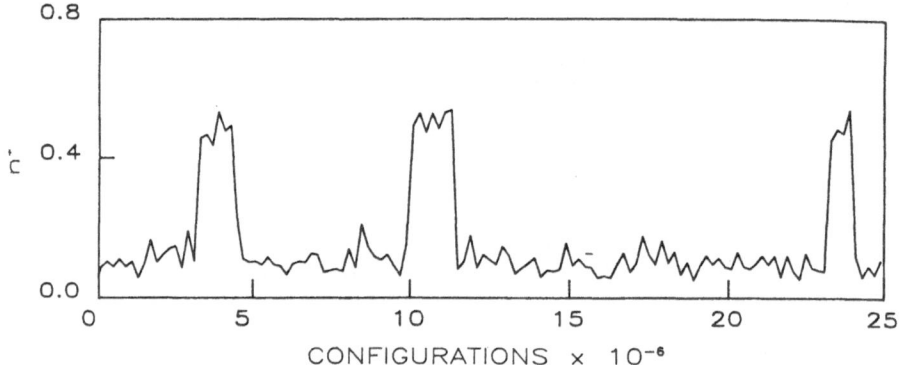

Figure 13. Same as Figure 11. s* = 7.9925, μ* = -12.20.

exhibits an increasingly sharp break, which appears to be discontinuous at h* = 4.9. Plots of local density on either side of the discontinuity indicate a vapor-liquid transition. If the area of the pore walls is too small, the fluid fluctuates (with the length of the Markov chain) between liquid and vapor phases over a range of values of μ* in the vicinity of the transition. That is, the isotherm exhibits hysteresis. By using larger-area walls we can narrow the region of hysteresis.

Following the interpretation of Evans et al. [55,56], we conclude that the apparent discontinuity in the isotherm for h* = 4.9 corresponds to the capillary-condensation transition, which occurs somewhere in the range -12.175 > μ* > -12.225. Moreover, the thermodynamic state of the pore fluid at h* = 4.9 must lie above the capillary critical point, whereas the thermodynamic states of the other two pores (h* = 3.05 and h* = 2.15) must lie in supercritical regions of their respective phase diagrams. We also note (see Figure 14) that the bulk condensation transition occurs at higher chemical potential than the corresponding capillary-condensation transition.

For the slit-pore at h* = 4.9, T* = 1.0 is below the critical temperature and therefore the slit-pore is in a region of its phase diagram similar to that of the smooth-walled cylindrical pore treated by Peterson and Gubbins [57], who found hysteresis in the isotherm at T* = 0.7. In order to narrow the hysteretic region of the slit-pore isotherm, we had to resort to longer GCEMC runs on a larger-area pore. A useful measure of the effort at exploring configuration space is the number of GCEMC steps per atom, which is of the order of 69 000 for points around the transition (μ* = -12.20, Figure 6). Peterson and Gubbins do not mention attempts to narrow their hysteretic region, which is based on GCEMC runs less than 9 million steps long for a pore 5σ in radius and 10σ in length. For a typical liquid density, their pore contains about 500 atoms, which corresponds to about 15 000 steps per atom.

From a more abstract viewpoint hysteresis in the context of the GCEMC method is the result of non-ergodic Markov chains. Wood [21] presents a very clear and useful discussion of ergodic problems associated with Monte Carlo techniques. In principle, the GCEMC Markov chain is ergodic. In practice, however, the system may

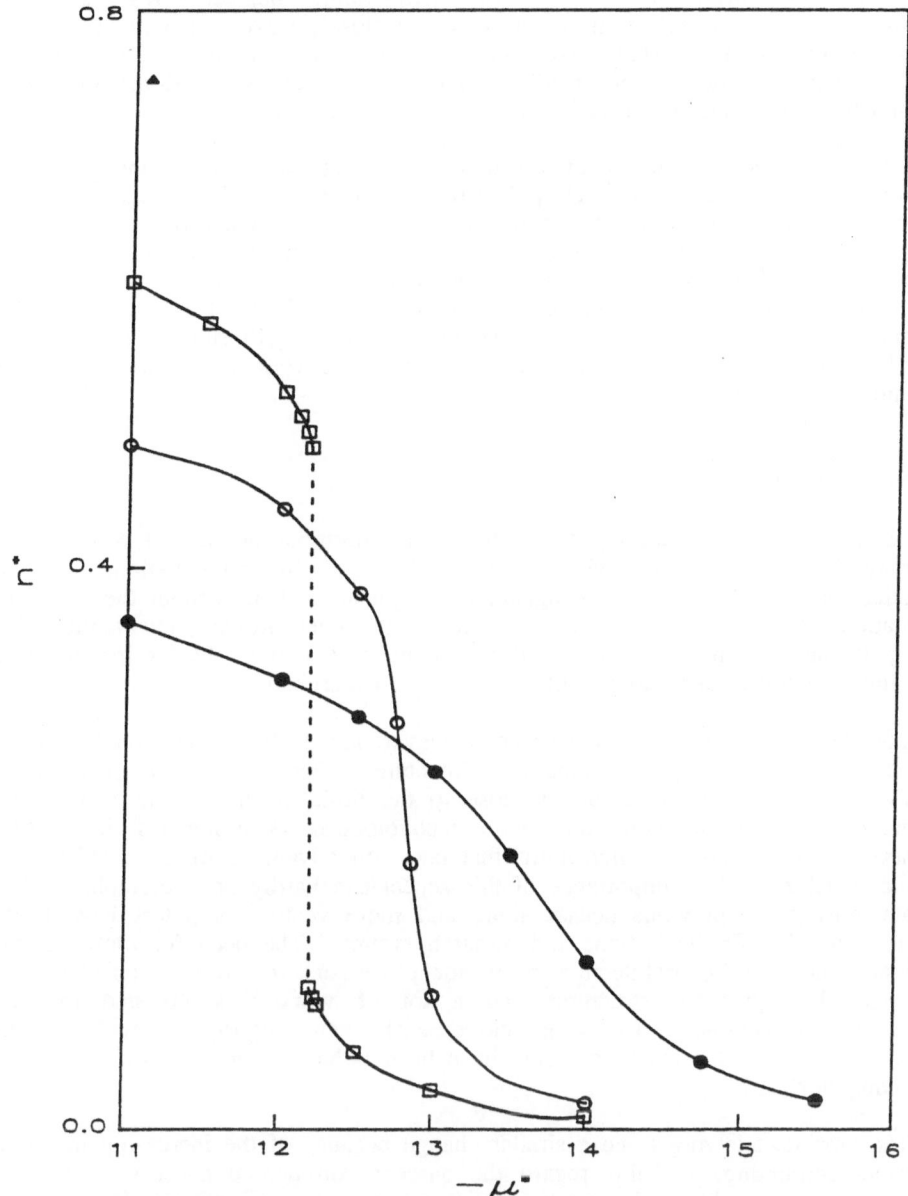

Figure 14. Sorption isotherms for pore at T* = 1.0.

(a) h* = 2.15 (●); (b) h* = 3.05 (○); (c) h* = 4.90 (□). Two points (▲) on the sorption isotherm of bulk fluid are also shown.

get trapped in a localized region of configuration space. In the number of steps available (dictated by one's computer budget usually) it may not be able to reach other regions that have significant probabilities of occurrence. Certain computed properties may be non-negligibly influenced by these inadvertently neglected regions. Wood states that the ergodic problem 'is not infrequently closely associated with the size and perhaps the shape of the system, and with the boundary conditions'. This comment is certainly apropos of microporous media because of the very small dimensions of the pores and because of the variation in the structure of their walls.

Hill [58] gives an excellent discussion of hysteresis within the context of equilibrium statistical mechanics. He points out that hysteresis cannot exist in a system truly at equilibrium; the system has a unique set of macroscopic properties in a given thermodynamic state. The appearance of hysteresis in equilibrium statistical mechanical theories is a result of various approximations to the true partition function (configuration integral) and not an exact result of physical law. This suggests that the appearance of hysteresis in experimental studies is a non-equilibrium phenomenon that cannot be properly understood within the restrictive framework of equilibrium statistical mechanics.

V. FINAL REMARKS

The hysteresis study clearly demonstrates the enormous amount of computer time that may be required to study certain phenomena by computational statistical mechanics. In fact, it is not an exaggeration to presume that, without the increase in computational speed over the years and without the ever growing availability of the most up to date machines, our knowledge of condensed matter physics in general and of the liquid state in particular would have remained marginal.

The increase in computational speed, however, had to be accompanied by several changes in the philosophy of computer architecture. This requires the scientist, who will always be working on problems close to the limit of the current generation of computers, to keep up constantly with technological developments in computer hardware and to design new algorithms that take full advantage of the architecture of particular machines. The importance of this aspect can hardly be overemphasized. For example, during the previous decade more and more vector computers have become accessible to scientific institutions and research groups. The need for more computer power and the specific architecture of vector processors forced a reconsideration of well established program structures and a lot of work was invested to design vectorizable algorithms. In some cases such new algorithms with optimum performance on a vector machine would have been rather inefficient on a conventional scalar computer [59].

The next decade might see a similar change because of the increasing importance of parallel computing. In this regard the nascent commercial success of transputer systems may help to boost this evolution. The very attractive price/performance ratio of transputer systems enables even smaller research groups to buy their own machines which can then be dedicated to the computational needs of just a few people in a flexible way. If we take into account turn-around-times, which can be quite large at poorly managed regional or university computer centers, transputer systems may actually compete with big mainframes in terms of the real time required for a calculation although, as we have shown here, even a large transputer system operates at only a fraction of the 'pure' computational speed of a vector machine [34].

ACKNOWLEDGEMENT

We are grateful to the computer center at Purdue University (PUCC) and to the scientific council of the Hchstleistungsrechenzentrum (HLRZ) at Forschungszentrum Jülich for generous allocations of computer time on the CYBER 205, the ETA 10-P (PUCC) and the CRAY-Y/MP 832 (HLRZ). We also thank Alfred Hertzner for the preparation of some of the figures. One of us (M.S.) is indebted to Professor Harald Morgner for his constant support.

REFERENCES

1. S.G. Brush, Kinetic Theory Vol. 1, in: *Selected Readings in Physics*, D. ter Haar, ed., Pergamon Press, Oxford (1965).
2. R. Clausius, *Sitz.-Ber. Niederrhein. Ges. Bonn*, 114 (1870).
3. J.C. Maxwell, *Philos. Mag.* 19:19 (1860), *ibid.* 20:21 (1860).
4. L. Boltzmann, *Sitz.-Ber. Akad. Wiss. Wien* (II) 66:275 (1872).
5. D.N. Subarev, *Statistische Thermodynamik des Nichtgleichgewichts*, Akademie Verlag, Berlin (1976).
6. J.P. Hansen and I.R. McDonald, *Theory of Simple Liquids* (2nd ed.), Academic Press (1986).
7. D.A. McQuarrie, *Statistical Mechanics*, Harper & Row, New York (1976).
8. J.G. Kirkwood, *J. Chem. Phys.* 3:300 (1935).
9. B.J. Alder, *Phys. Rev. Lett.* 12:317 (1964).
10. G.H.A. Cole, *Adv. Phys.* 8:225 (1959) and *J. Chem. Phys.* 28:912 (1958).
11. M.P. Allen, and D.J. Tildesley, *Computer Simulation of Liquids*, Clarendon Press, Oxford (1987)
12. M.H. Kalos, ed., *Monte Carlo Methods in Quantum Problems*, D. Reidel Publishing Co., Dordrecht (1984).
    M. Suzuki, ed., *Quantum Monte Carlo Methods in Equilibrium and Nonequilibrium Systems*, Springer, Heidelberg (1987).
13. W.W. Wood and J.D. Jacobson, *J. Chem. Phys.* 27:1207 (1957) and W.W. Wood and F.R. Parker, *ibid.*, 27:720 (1957).
14. B.J. Alder, and T.E. Wainwright, *J. Chem. Phys.* 27:1208 (1957).
15. C.L. Brooks, M. Karplus and B.M. Pettitt, *Adv. Chem. Phys.* Vol. LXXI, John Wiley & Sons, New York (1988).
16. G.R. Luckhurst, and G.W. Gray, *The Molecular Physics of Liquid Crystals*, Academic Press, New York (1979).
17. F.F. Abraham, W.E. Rudge, D.J. Auerbach and S.W. Koch, *Phys. Rev. Lett.* 52:445, (1984).
18. J.S. Rowlinson, and B. Widom, *Molecular Theory of Capillarity*, Clarendon Press, Oxford (1982).
19. D. Nicholson, and N. Parsonage, *Computer Simulation and the Statistical Mechanics of Adsorption*, Academic Press, New York (1982).
20. N. Metropolis, and S. Ulam, *J. Am. Stat. Assoc.* 44:335 (1949).
21. W.W. Wood, in: *Physics of Simple Liquids*, H.N.V. Temperley, J.S. Rowlinson and G.S. Rushbrooke, eds., North Holland, Amsterdam (1968).
22. J.P. Valleau, and S.G. Whittington, in: *Statistical Mechanics A. Modern Theoretical Chemistry*, B.J. Berne, ed., Plenum Press, New York (1977).
23. K. Binder, ed., *Monte Carlo Methods in Statistical Physics* (2nd ed.), Springer, Berlin (1986).
24. K.L. Chung, *Markov Chains with Stationary State Probabilities* Vol. 1, Springer, Heidelberg (1960).

25. N. Metropolis, A.W. Rosenbluth, M.N. Rosenbluth, A.H. Teller and E. Teller, *J. Chem. Phys.* 21:1087 (1953).
26. W.W. Wood, in: *Proceedings of the Enrico Fermi Summer School*, G. Ciccotti and W.G. Hoover, eds., Varenna (1985).
27. M. Parrinello and A. Rahman, *Phys. Rev. Lett.* 45:1196 (1980).
28. M. Parrinello and A. Rahman, *J. Appl. Phys.* 52:7182 (1981).
29. M. Parrinello and A. Rahman, *J. Chem. Phys.* 76:2662 (1982).
30. J. Boris, *J. Comput. Phys.* 66:1 (1986).
31. M. Schoen, *Comput. Phys. Commun.* 52:175 (1989).
32. G.S. Grest, B. Dünweg and K. Kremer, preprint 1989.
33. R. Vogelsang, in: *Proceedings of the 1984 Conferences on CYBER 200 in Bochum*, H. Ehlich, K. Schloer and B. Wojcieszynski, eds., Bochum (1985).
34. M. Luckas, *Scientific Computing & Automation*, 57 (1989).
35. J.L. Lebowitz, J.K. Percus and L. Verlet, *Phys. Rev.* 153:250 (1967).
36. B. Widom, *J. Chem. Phys.* 86:869 (1982).
37. R.E. Grim, *Clay Minerology* (2nd ed.), McGraw-Hill (New York, 1968).
38. U. Heinbuch, and J. Fischer, *Molec. Simulation* 1:109 (1987).
39. R.M. Pashley, and J.M. Israelachvili, *J. Coll. Interface Sci.* 101:511 (1984).
40. J.H. Cushman, D.J. Diestler and M. Schoen, to be published (1990).
41. I.K. Snook and W. van Megen, J. Chem. Phys. 72:2907 (1980).
42. W.A. Steele, *Interaction of Gases with Solid Surfaces*, Pergamon Press, Oxford (1974).
43. J.E. Lane and T.H. Spurling, *Chem. Phys. Letters* 67:107 (1979).
44. J.J. Magda, M. Tirrell and H.T. Davis, *J. Chem. Phys.* 83:1888 (1985);
45. M. Schoen, D.J. Diestler and J.H. Cushman, *J. Chem. Phys.* 87:5464 (1987).
46. S. Toxvaerd, *J. Chem. Phys.* 74:1998 (1981).
47. C.L. Rhykerd Jr., M. Schoen, D.J. Diestler and J.H. Cushman, *Nature* 330:461 (1987).
48. M. Schoen, J.H. Cushman, D.J. Diestler, and C.L. Rhykerd Jr., *J. Chem. Phys.* 88:1394 (1988).
49. R. Steininger and J. Bilgram, *Helvetica Physica Acta* 62:215 (1989). See also: P. Böni, J.H. Bilgram and W. Känzig, *Phys. Rev.* A28:2953 (1983); U. Dürig, J.H. Bilgram and W. Känzig, *Phys. Rev.* A30:946 (1984); and P.U. Halter, J.H. Bilgram and W. Känzig, *J. Chem. Phys.* 89:2622 (1988).
50. L.A. Rowley, D. Nicholson and N.G. Parsonage, *J. Comput. Phys.* 26:66 (1975).
51. J. Yao, R.A. Greenkorn and K.C. Chao, *Mol. Phys.* 46:587 (1982).
52. G.E. Norman and V.S. Filipov, *High Temp. (USSR)* 7:216 (1969).
53. D.J. Adams, *Mol. Phys.* 29:307 (1975).
54. M. Schoen, C.L. Rhykerd Jr., J.H. Cushman and D.J. Diestler, *Mol. Phys.* 66:1171 (1989).
55. R. Evans, U. Marini Bettolo Marconi and P. Tarazona, *J. Chem. Phys.* 84:2377 (1986).
56. R. Evans, U. Marini Bettolo Marconi and P. Tarazona, *J. Chem. Soc. Faraday II* 82:1763 (1986). See also: J.P.R.B. Walton and N. Quirke, *Molec. Simulation* 2:361 (1989).
57. B.K. Peterson and K.E. Gubbins, *Mol. Phys.* 62:215 (1987).
58. T. Hill, *Statistical Mechanics*, McGraw-Hill, New York (1956), p.164.
59. U. Meier, M. Schindler and V. Staemmler, in: *Proceedings of the 1985 Conferences on Supercomputers and Applications*, Bochum (1986).

THE USEFULNESS OF VECTOR COMPUTERS FOR PERFORMING

SIMULTANEOUS EXPERIMENTS

Pieter Moerman

Laboratorium voor Informatica, State University of Ghent (RUG)
Grotesteenweg-Noord 2, B-9710 Gent-Zwijnaarde, Belgium

1. BASIC PRINCIPLES

Computers are often used in theoretical science for simulations of the behaviour of systems. The vectorization of the programs can however in many cases be difficult because of the sequential nature of the experiment. In some of these cases, we can still use vector computers, by introducing a new approach to the solution of the problem, for example using vectors to represent multiple simultaneous executions of the program.

In many of the problems in fields where probability plays a part, for instance in quantum mechanics, the execution of the program depends for a great deal on random numbers. When we want to use computers to simulate these problems, it may be necessary to consider multiple runs of the program to discard all effects due to the randomness in the problem to get a reliable view. Even when the problem at hand is hardly vectorizable, we can often use a vector computer in such cases to perform several simultaneous experiments in one run of the program. We represent the quantities as vectors instead of scalars, so that each element of the vector represents the quantity in one experiment.

In the following paragraphs, we will illustrate this concept with a few examples. The first consists of a simulation of a number of throws with a die. We have chosen this example for its simplicity; no prior knowledge in a specific field is needed.

The second example describes a specific problem taken from the field of quantum mechanics: the calculation of path integrals.

All the programs in this text are written in Cyber 200 Fortran which allows explicit vectorization (see [1]).

2. EXAMPLE: THROWING DICE

We will first consider a simple example. We will simulate ten thousand throws with a die. (This can be seen as a simple test of the random generator of our system.) This is easy to achieve: we simple generate ten thousand integer random numbers between one and six (see program 1 in Figure 1). This program is not vectorizable in a trivial way.

program 1

```
          PROGRAM DICE
          PARAMETER (NTHROWS=10000)
          DIMENSION ICOUNT(6)
          DO 5 I=1,6
            ICOUNT(I)=0
     5    CONTINUE
          DO 10 K=1,NTHROWS
            ITHROW=INT(RANF()*6.)+1
            ICOUNT(ITHROW)=ICOUNT(ITHROW)+1
    10    CONTINUE
          PRINT 100,(ICOUNT(I),I=1,6)
   100    FORMAT (6I5)
          STOP
          END
```

Figure 1

program 2

```
          PROGRAM DICE2
          PARAMETER (NTHROWS=10000,NSIMUL=1000)
          DIMENSION ICOUNT(6,NSIMUL),ITOTAL(6)
          DO 30 J=1,NSIMUL
            DO 10 I=1,6
              ICOUNT(I,J)=0
    10      CONTINUE
            DO 20 K=1,NTHROWS
              ITHROW=INT(RANF()*6.)+1
              ICOUNT(ITHROW,J)=ICOUNT(ITHROW,J)+1
    20      CONTINUE
    30    CONTINUE
          DO 50 I=1,6
            ITOTAL(I)=0
            DO 40 J=1,NSIMUL
              ITOTAL(I)=ITOTAL(I)+ICOUNT(I,J)
    40      CONTINUE
    50    CONTINUE
          PRINT 100,(ITOTAL(I),I=1,6)
   100    FORMAT (6I5)
          STOP
          END
```

Figure 2

```
        PROGRAM DICE3
        PARAMETER (NTHROWS=10000,NSIMUL=1000)
        DIMENSION ITOTAL(6)
        DESCRIPTOR IVTHROW,VTHROW,ICOUNT(6)
        ASSIGN VTHROW,.DYN.NSIMUL
        ASSIGN IVTHROW,.DYN.NSIMUL
        DO 5 I=1,6
          ASSIGN ICOUNT(I),.DYN.NSIMUL
          ICOUNT(I)=0
   5    CONTINUE
        DO 20 K=1,NTHROWS
          CALL VRANF(VTHROW,NSIMUL)
          VTHROW=VTHROW*6.
          IVTHROW=VINT(VTHROW;NSIMUL)+1
          DO 10 I=1,6
          WHERE(IVTHROW.EQ.I)ICOUNT(I)=ICOUNT(I)+1
  10      CONTINUE
  20    CONTINUE
        DO 30 I=1,6
          ITOTAL(I)=Q8SSUM(ICOUNT(I))
  30    CONTINUE
        PRINT 100,(ITOTAL(I),I=1,6)
 100    FORMAT (6I7)
        STOP
        END
```

Figure 3

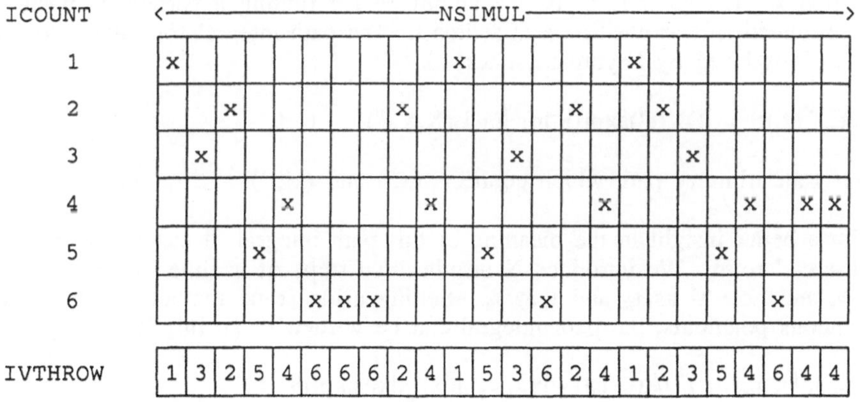

Figure 4

Now, we are going to extend our simulation, we will perform a thousand simulations of the experiment described above. Again, this is fairly easy to program on a sequential computer by nesting two loops (see program 2 in Figure 2). We can vectorize this using the technique described in the first paragraph: we use vectorization to perform the simulations simultaneously.

Instead of generating successive random numbers, we generate random vectors with a length of NSIMUL; each element of the vector corresponds to a result in one simulation. The results are accumulated in the vectors ICOUNT(I). At the end of the program, we can calculate the total frequency by summing the elements of the vectors (see program 3 in Figure 3).

Generating the random vector corresponds to simultaneously throwing NSIMUL dice; each element of the vector corresponds to a throw in one simulation. The vectors ICOUNT(I) count the frequency of the number I in all the simulations. Here we have a little problem; only those elements of ICOUNT(I) for which the corresponding element of IVTHROW is equal to I must be increased (see also Figure 4). We solve this problem by using the bit vector IVTHROW.EQ.I to mask the elements which must be increased, in combination with a WHERE instruction. The WHERE instruction is the vector equivalent of the IF instruction. It performs the subsequent vector operations only on those elements of the vectors for which the corresponding bit in the bit vector is equal to 1. The calculation of the total frequency is performed by summing the elements in the vectors of ICOUNT.

3. EXAMPLE: PATH INTEGRALS

We will consider the calculation of path integrals as a second example. For a more elaborate description of path integrals and their applications in quantum field theory, we refer to specialized literature (see for example [2] or [3]). Here, we will give a brief survey of the problem and a practical method to solve such problems.

a) Outline of Path Integral Formulation of Quantum Mechanics

Consider a one-dimensional system, described by the Lagrange function

$$L(\dot{x},x,t) = (m/2)\dot{x}^2 - V(x) \tag{1}$$

where $\dot{x}$ is the velocity dx/dt, m the mass and V(x) the potential energy of a particle at the position x. Let $<x',t'|x,t>$ be the amplitude for finding a particle at position x' at time t', provided it is at position x at a time t. The path integral formulation of quantum theory, as introduced by Feynman, states:

$$<x',t'|x,t> = \int DX(t'')\exp[(i/\hbar)\int dt''L(\dot{X},X,t'')] \tag{2}$$

with X(t'') an arbitrary path which connects (x,t) and (x',t').

For a better insight in the meaning of this path integral, it can be better to write it in a 'discrete' form. We introduce N infinitesimal steps Δt in time, so that $\Delta t = (t'-t)/N$, $N \to \infty$, and note t' as t_N and t as t_0. Setting aside some mathematical problems for discontinuous potentials, the path integral can be written as an infinite multiple integral:

$$<x_N,t_N|x_0,t_0> = (1/A(t_N-t_0)) \int dx_{N-1} \dots \int dx_2 \int dx_1 \exp[iS/\hbar]$$

$$S = \sum \Delta t[(m/2)\ (x_{j+1}-x_j)/\Delta t)^2 - V(x_j)] \tag{3}$$

where the normalization factor is known, but will be treated later.

For more elaborate derivations, explanations and interpretations, we refer to specialized literature.

Now we will concentrate on the wave function. Let $\Psi(x_0,t_0) = \langle x_0,t_0|\Psi\rangle$ be the wave function in position x_0 at a time t_0, then the wave function $\Psi(x_N,t_N)$ in x_N at a time t_N is given by

$$\Psi(x_N,t_N) = \int dx_0 \langle x_N,t_N|x_0,t_0\rangle \Psi(x_0,t_0) \tag{4}$$

so the calculation of the path integral gives us the solution of the time-dependent Schrödinger equation given the starting condition $\Psi(x_0,t_0)$.

On the other hand, it is well known that the wave function $\Psi(x,t)$ can be written as a linear combination of the eigenfunctions $\psi_n(x)$ of the Hamilton function related to the energy eigenvalue E_n:

$$\Psi(x,t) = \sum c_n(t_0)\ \psi_n(x)\ \exp[-iE_n(t-t_0)/\hbar] \tag{5a}$$

where the expansion coefficients are determined by the starting condition:

$$c_n(t_0) = \int dx_0\ \psi_n^*(x_0)\ \psi_n(x_0,t_0) \tag{5b}$$

This means that the amplitude $\langle x_N,t_N|x_0,t_0\rangle$ can be written in terms of the stationary eigenfunctions $\psi_n(x)$:

$$\langle x_N,t_N|x_0,t_0\rangle = \sum \psi_n^*(x_0)\ \psi_n(x_N)\ \exp[-iE_n(t_N-t_0)/\hbar] \tag{6}$$

This development is very illustrative if we consider it at 'imaginary time'.
Let $t_N-t_0 = i\tau$, and choose without loss of generality $t_0 = 0$. Let $\Delta t = -i\varepsilon$. Then:

$$\langle x_N,-i\tau|x_0,0\rangle = \sum \psi_n^*(x_0)\ \psi_n(x_N)\ \exp[-E_n\tau/\hbar]$$

$$= (1/A(-i\tau)) \int dx_{N-1} \ ... \ \int dx_1\ \exp[-\varepsilon H/\hbar] \tag{7}$$

$$H = \sum [(m/2)\ ((x_{j+1}-x_j)/\varepsilon)^2 + V(x_j)]$$

If the ground state is not degenerate, the particles density $|\psi_0(x)|^2$ of the ground state can easily be determined. If we take equations (6) and (7) for $x_N = x_0$, we get:

$$\sum |\psi_n(x_0)|^2\ \exp[-E_n\tau/\hbar] \tag{8}$$

$$= (1/A(-i\tau)) \int dx_N \int dx_{N-1} \ ... \ \int dx_1 \delta(x_N-x_0)\ \exp[-\varepsilon H/\hbar]$$

where the closed path was expressed by the δ-function. If we consider the case $\tau \gg \hbar/(E_0 -E_1)$ (which we will consider for the rest of this paper), only the ground state contributes to the left hand side of equation (8), so

$$|\psi_0(x_0)|^2 = (\exp[E_0\tau/\hbar]/A(-i\tau)) \int dx_N \ ... \ \int dx_1 \delta(x_N-x_0)\ \exp[-\varepsilon H/\hbar]$$

If we normalize the wave function, this equation can be written as

$$|\psi_0(x_0)|^2 = \int dx_N \ ... \ \int dx_1 \int dx_0 \delta(x-x_0)\ (\delta(x_N-x_0)\ \exp[-\varepsilon H/\hbar]/Z) \tag{9}$$

with

$$Z = \int dx_N \ ... \ \int dx_1 \int dx_0 \delta(x_N-x_0)\ \exp[-\varepsilon H/\hbar] \tag{10}$$

b) Application of the Monte Carlo and Metropolis Technique

In equation (9), the wave function is expressed as an infinite multiple integral of the form:

$$|\psi_0(x)|^2 = \int d^N r\delta(x-x_0)w(\hat{r}); \quad \hat{r} = (x_0, x_1, ..., x_N) \tag{11a}$$

where the weight function $w(\hat{r})$ is of the form:

$$w(\hat{r}) = (1/Z)\ \exp[-\epsilon H(x_0, x_1, ..., x_N)/\hbar] \tag{11b}$$

For multiple integrations, the Monte Carlo method in spite of its slowness, is often the best numerical method:

$$\int d^N r f(\hat{r}) \equiv V\!\!<\!\!f\!\!> \pm V\sqrt{(<\!\!f^2\!\!>-<\!\!f\!\!>^2)}/N$$

$$<\!\!f\!\!> = (1/N)\ f(\hat{r}_i)$$

$$<\!\!f^2\!\!> = (1/N)\ f^2(\hat{r}_i)$$

where N points $\hat{r}_1, \hat{r}_2, ..., \hat{r}_N$ are chosen randomly from a uniform distribution over the multidimensional volume V. The weight function $w(\hat{r})$ gives in this problem a serious complication due to the normalization factor Z. The 'Metropolis' algorithm (see also [4] for a more explicit description) gives us a solution for this problem. This procedure lets us choose the points randomly from a normalized probability distribution $w(\hat{r})$, so that

$$\int d^N r f(\hat{r})w(\hat{r}) \equiv (1/N) \sum f(\hat{R}_i)$$

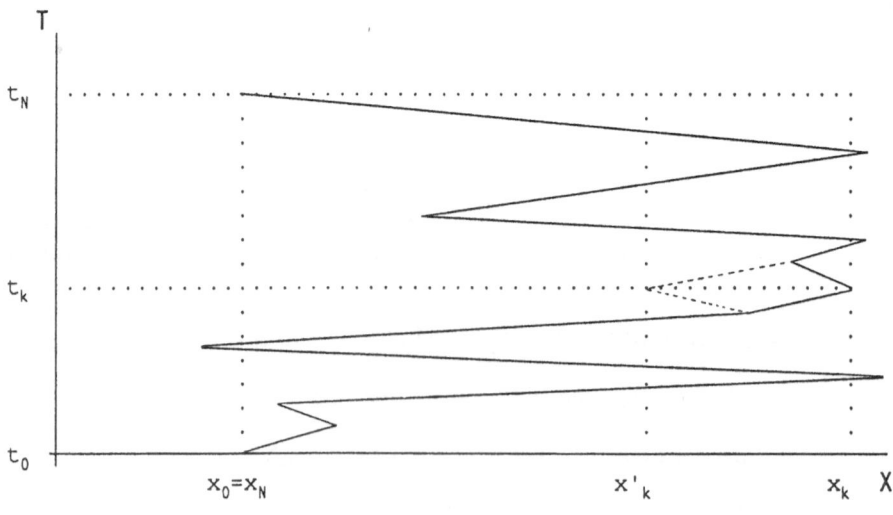

Figure 5

where $\hat{R}_1, \hat{R}_2, ..., \hat{R}_0$ are distributed according to $w(\hat{R})$. The Metropolis algorithm is in this case very simple. Consider a randomly chosen path from a point x_0 at a time t_0 to a point $x_N = x_0$ at a time t_N. If we change a randomly chosen point x_k on this path to a point x'_k (see Figure 5), then we have a change of

$$\Delta H = H(x_0,...,x'_k,...,x_N) - H(x_0,...,x_k,...,x_N)$$

in the argument of the exponent of the weight function $w(\hat{r})$. The Metropolis algorithm tells us to accept this point if $\Delta H < 0$, or if $\exp[-\epsilon\Delta H/\hbar] > \xi$, where ξ is a randomly chosen number from a uniform distribution, $0 \le \xi < 1$. If the point x'_k is accepted according to this rule, the point x_k is changed to x'_k, and the path $x_0,...,x_k=x'_k,...,x_N=x_0$ belongs to the randomly chosen paths determined by $w(\hat{r})$. This new point x_k adds to the probability distribution in x_k, i.e. $|\psi_0(x_k)|^2$ must be increased to $|\psi_0(x_k)|^2+1$. The other points of the distribution satisfy also the distribution, so the probability distribution in these points must also be increased.

c) Sequential Program

First we give a formal (sequential) solution to the problem. We do not intend to give here a full program which covers all cases. We just want to indicate how such a program can be written and how it can be vectorized. The interested reader can adapt the code to his needs and wishes.

The parameters we use in the algorithm are:
NPTS: number of intervals on the x-axis
XMIN: beginning point of the interval on the x-axis
XMAX: ending point of the interval on the x-axis
NTIME: number of intervals on the time-axis
EPS1, EPS2: the two values which are used in the subroutine (see later)
AC: the accuracy used (see later)

The wave function is stored in an array PSI. This array has NPTS+2 elements: PSI(0)...PSI(NPTS-1) gives the wave function in the first, second, ..., the last of the NPTS intervals; in PSI(-1) and PSI(NPTS) we accumulate the points respectively left and right of the interval.

The array IX contains the successive points of the path. We have introduced a complete discretization, so that IX does not represent the position of the particle, but the subinterval in which this position is. We take the beginning point of this interval as the value.

<u>program 4/1</u>

```
      PARAMETER (NPTS=512,NTIME=512,XMIN=-4.0,XMAX=4.0,
     X                           EPS1=300.0,EPS2=0.05,AC=0.10)
      DIMENSION IX(NTIME),PSI(-1:NPTS)
```

Figure 6

```
        T=1/EPS1
        DO 80 NPASS=1,2
          IF (NPASS.EQ.2) T=1/EPS2
30      NSUCC=0
        DO 50 ICOUNT=1,NPTS*100
          K=INT(NTIME*RANF())+1
          IDELTA=1
          IF (RANF().LT.0.5) IDELTA=-1
          IXK=IX(K)+IDELTA
          ...
          DE=((XK-X)*T)*((XK+X-XP-XM)*T)+(V(XK)-V(X))
          CALL METROP (DE,T,ANS)
          IF (ANS) THEN
            NSUCC=NSUCC+1
            IF (NPASS.EQ.2) THEN
              DO 40 I=1,NTIME
                IF (I.EQ.K) THEN
                  J=IXK
                ELSE
                  J=IX(I)
                ENDIF
                PSI(J)=PSI(J)+1.
40            CONTINUE
            ENDIF
            IF ((IXK.GE.0).AND.(IXK.LT.NPTS)) IX(K)=IXK
          ENDIF
          IF (NSUCC.EQ.10*NPTS) GOTO 60
50      CONTINUE
60      ...
        EQUILIB=MAX(ABS((AV-AV2)/AV),ABS((VAR-VAR2)/VAR)).LE.AC
        ...
        IF (.NOT.EQUILIB) GOTO 30
80      CONTINUE
```

Figure 7

The algorithm is executed twice with a different value of ε; first a relatively high value, at the second passage a relatively small value. These two values are determined by the user. The iteration comes to an end if the path does not change much anymore (a relative change in the mean and the variance of the path not greater than AC). This condition is checked after 10*NPTS successful changes in the path, or after 100*NPTS attempts to change (whatever comes first). Only the changes in the second iteration are calculated in PSI. The first iteration serves to get the system close to its ground state.

We first choose a time interval at random (K). Furthermore, we only consider changes in the path to adjacent intervals (IDELTA equal to 1 or -1 respectively for a change to the right or the left). This way we have a bigger chance of finding the right path more quickly.

```
      SUBROUTINE METROP (DE,T,ANS)
C METROPOLIS ALGORITHM
      LOGICAL ANS
      ANS=(DE.LT.0.).OR.(RANF().LT.EXP(-DE/T))
      RETURN
      END
```

Figure 8

To calculate the change in the energy of the system, we only have to take the old and the new position of the particle into account, and the position of the particle before and after our randomly chosen moment (respectively X, XK, XM and XP). The function V is the potential energy.

The subroutine METROP is the implementation of the Metropolis algorithm.

It can happen that the algorithm approves of a change to a point outside the given interval between XMIN and XMAX, though this should seldom occur. These points are accumulated in PSI(-1) and PSI(NPTS) respectively for the points left and right of the interval. Of course the changes are not executed in such cases.

At the end of the program, we still have to normalize the wave function PSI.

d) Vectorization of the Program

We vectorize the program in the way we described at the beginning of this text. We perform simultaneous calculations of the path integral; then we calculate a wave function from all the results.

The vectorized program is on the whole not very different from the sequential program. We add one parameter, NTESTS, which is the number of tests the user wants to perform simultaneously. One execution of the vectorized program with parameter NTESTS is equivalent to NTESTS executions of the scalar program, with a processing of the results.

program 5/1

```
   PARAMETER (NPTS=512,NTIME=512,XMIN=-4.0,XMAX=4.0,
  X               EPS1=300.0,EPS2=0.05,AC=0.10,NTESTS=1000)
   DIMENSION PSI(-1:NPTS)
   DESCRIPTOR IX(NTIME),VPSI(-1:NPTS),AV,VAR,X,NSUCC,IXK,XK,XP,XM,DE,
  X                    ANS,J,IDELTA,VTEMP,VTEMP2
   BIT ANS
```

Figure 9

237

```
          T=1/EPS1
          DO 80 NPASS=1,2
            IF (NPASS.EQ.2) T=1/EPS2
      30    NSUCC=0
            DO 50 ICOUNT=1,NPTS*100
              K=INT(NTIME*RANF())+1
              IDELTA=1
              CALL VRANF (VTEMP,NTESTS)
              WHERE (VTEMP.LT.0.5) IDELTA=-1
              IXK=IX(K)+IDELTA
              ...
              VTEMP=V(XK;VTEMP)
              VTEMP2=V(X;VTEMP2)
              DE=((XK-X)*T)*((XK+X-XP-XM)*T)+(VTEMP-VTEMP2)
              CALL VMETROP (DE,T,ANS,NTESTS)
              WHERE (ANS) NSUCC=NSUCC+1
              IF (NPASS.EQ.2) THEN
                DO 40 I=1,NTIME
                  IF (I.EQ.K) THEN
                     J=IXK
                  ELSE
                     J=IX(I)
                  ENDIF
                  DO 35 L=-1,NPTS
                    WHERE (ANS.AND.(J.EQ.L)) VPSI(L)=VPSI(L)+1.
      35          CONTINUE
      40        CONTINUE
              ENDIF
              WHERE (ANS.AND.(IXK.GE.0).AND.(IXK.LT.NPTS)) IX(K)=IXK
              NMAX=Q8SMAX(NSUCC)
              IF (NMAX.GE.10*NPTS) GOTO 60
      50    CONTINUE
      60    ...
          EQUILIB=MAX(AVMAX,VARMAX).LE.AC
          ...
          IF (.NOT.EQUILIB) GOTO 30
      80 CONTINUE
```

Figure 10

The vectorization of the algorithm is rather straightforward: most of the variables are substituted by vectors; the operations on these variables become simultaneous operations in all tests. The vector array IX for example gives the position of the particle in each test. In the same way the vector array VPSI describes the corresponding wave functions.

We choose the same point in each of the tests, in order not to make the program too complicated. This point is changed for each of the tests, then for each test the difference DE in energy is calculated and the Metropolis algorithm is used. Because of

```
      SUBROUTINE VMETROP (DE,T,ANS,ILENGTH)
C METROPOLIS ALGORITHM
      DESCRIPTOR DE,ANS,VTEMP1,VTEMP2
      BIT ANS
      ...
      VTEMP1=-DE/T
      VTEMP1=VEXP(VTEMP1;VTEMP1)
      CALL VRANF (VTEMP2,ILENGTH)
      ANS=(DE.LT.0.).OR.(VTEMP2.LT.VTEMP1)
      RETURN
      END
```

Figure 11

the fact that in Cyber 200 Fortran, we cannot use vector functions directly in vectorial expressions, we have to introduce some new vector variables for calculating purposes (VTEMP and VTEMP2).

The calculation of the wave function is performed in the same way as the calculation of the frequency in the first example.

After 100*NPTS attempts or if at least one of the tests has produced 10*NPTS successful attempts, the program checks if the equilibrium is reached. We repeat this procedure until all systems in all tests have reached their state of equilibrium.

The subroutine VMETROP is the vectorized equivalent of METROP. It produces a bit vector which indicates for each simultaneous calculation whether or not the change in that calculation should be accepted.

At the end of the calculation, all the wave functions are normalized. Then the final wave function is derived from these functions. We could for example take the mean value in each point of all wave functions; however, another choice is very well possible. Note that, since the shape of the wave function can differ with the application, and in most of the cases is not known in advance, we generally cannot use a least squares approximation. However, since the wave functions should not be very different from each other, the mean value should do well.

It is difficult to compare the performance of the two programs since the time of execution depends largely on random numbers. However, there is a considerable gain in ease of use: one execution of the vectorized program is equivalent to a number of executions of the scalar program, with a further processing of the results. In this way we can perform up to 65000 calculations of one path integral simultaneously.

CONCLUSIONS

Vector computers can be very useful in theoretical science. The examples given in the previous paragraphs show that the use even can be extended to applications which seem hardly vectorizable, by using a new approach to the problem. Therefore, the problems for which vector computers are to be used, should be studied carefully so as to take a maximal advantage of the facilities of the machine.

ACKNOWLEDGMENT

We wish to thank Prof. Dr. J.T. Devreese and Dr. F. Brosens of the University of Antwerp (UIA) for their cooperation in the development of the application described in paragraph 3 (calculation of path integrals).

The calculations could be carried out thanks to the material support of the NFSR.

REFERENCES

1. *CDC Cyber 200 Fortran Version 2 Reference Manual*, Control Data Corporation, Sunnyvale (1981).
2. R.P. Feynman and A.R. Hibbs, *Quantum Mechanics and Path Integrals*, McGraw Hill, New York (1965).
3. L.S. Schulman, *Techniques and Applications of Path Integration*, Wiley Interscience, New York (1981).
4. W.H. Press et al., *Numerical Recipes: The Art of Scientific Computing*, Cambridge University Press, Cambridge (1986).

AUTHOR INDEX

Abraham, F.F., 197, 227
Adams, D.J., 217, 228
Agoshkov, V.I., 145, 158
Ahlrichs, R., 123, 127, 132, 134, 136, 137
Akai, H., 90, 96
Alder, B.J., 197, 227
Allan, R., 134, 137
Allen, M.P., 197-201, 203-205, 208, 211, 227
Almlöf, J., 127, 137
Anderson, D.V., 52, 66, 81, 159, 164, 173
Auerbach, D.J., 197, 227
Auger, J.M., 193
Aziz, A.K., 147, 158

Bachelet, G., 94, 96
Badziag, F., 90, 96
Bär, M., 132, 136, 137
Benner, R.E., 193
Berger, D., 160, 173
Bernard, L.C., 160, 173
Berne, B.J., 198, 203, 227
Bilgram, J., 228
Bilgram, J.H., 216, 228
Binder, K., 198, 203, 227
Binkley, J.S., 123, 136
Birch, F., 96
Bjordstad, P.E., 144, 158
Böhm, H.-J., 123, 136
Boltzmann, L., 195, 227
Böni, P., 216, 228
Bonomi, E., 51, 52, 54, 61, 81
Boris, J., 209, 228
Bossel, U.G., 52, 68, 82
Bourgat, J.F., 146, 158
Boyle, J., 84, 96
Brebbia, C.A., 101, 121
Brezzi, F., 147, 158
Brooks, C.L., 197, 227
Brosens, F., 23, 24, 50, 95, 96

Brush, S.G., 195, 227
Brüstel, U., 85, 96
Buell, D.A., 124, 136
Burkhardt, A., 123, 130, 136

Carriero, N., 194
Chan, T.F., 142, 146, 157, 158
Chao, K.C., 217, 228
Chin, S., 123, 136
Chung, K.L., 202, 203, 227
Ciccotti, G., 205, 228
Clausius, R., 195, 227
Clementi, E., 123, 136
Cole, G.H.A., 197, 227
Colvin, M.E., 123, 136
Cooper, W.A., 52, 66, 81, 159, 164, 173
Corongiu, G., 123, 136
Cremer, D., 127, 137
Cushman, J.H., 195, 213, 215, 216, 218, 220, 228

Davidson, E.R., 129, 132, 137
Davidson, E.S., 103, 121
Davis, H.T., 213, 228
De Roeck, Y.H., 146, 158
Dederichs, P., 89, 90, 96
Degtyarev, L.M., 160, 173
Dekker, K., 17, 21
Denteneer, P.J., 95, 96
Desbiolles, J.J., 52, 66, 82
Detrich, J., 123, 136
Devreese, J.T., 23, 24, 50, 83, 84, 91, 93, 95-97
Dewar, D.L., 163
Dewar, M.J., 90, 96
Dewar, R.L., 160, 173
Diestler, D.J., 195, 213, 215, 216, 218, 220, 228
Domingo, L., 123, 136
Dongarra, J.J., 20, 21, 84, 96, 102, 121
Dorizzi, B., 193
Droux, J.J., 52, 66, 82

SUBJECT INDEX

Ada, 178
Addressing, 107
Amdahl's law, 10
Assembly line principle, 1

Bank stack, 116, 117
Bank-busy-time, 8, 9
BBGKY hierarchy, 196
BGY hierarchy, 197
Bisection method, 84
BLAS, 20, 192
Block-Gaussian elimination, 141
Boltzmann's equation, 196
Bulk modulus, 91

C, 178-182, 189
Canonical ensemble, 203
CEDAR, 103
Chaining, 4, 5
Chebychev polynomial, 47
Choleski factorization, 155, 156
Comparex, 132
Configuration interaction, 123
Conjugate gradient method, 63, 139, 145, 154, 156
Connection Machine, 166
Continuous Pipe Vector Computer, 99, 103
CONVEX, 4, 8, 108, 120
CRAY X-MP, 4-6, 87, 103, 104, 110, 120, 132
CRAY Y-MP, 4, 9, 19, 103, 104, 120, 155, 159, 160, 172, 210
CRAY-1, 2, 101, 105, 108
CRAY-2, 9, 19, 53, 63, 102-104, 120, 155, 160, 172
CRAY-3, 102
CRAY-4, 102, 105
CSTools, 179
Cubix, 186

CYBER 205, 3, 19, 24, 50, 84, 87, 101, 209, 210
Cyclic reduction, 17

Data structure, 11, 13, 16
Data transfer, 101, 115
Data-flow computer, 119
Dense fluid, 196
Density functional theory, 83, 93, 123
Dirichlet boundary conditions, 15, 141, 145, 146
Divide and conquer, 17
Domain decomposition method, 157
Dyadic operations, 6

Eigenvalue problem, 83
Elliptic partial differential equation, 15, 139, 140
ETA 10, 3, 24, 28, 32, 41, 50, 84, 87, 101, 198, 209
Euler flow, 60
Exclusive or method, 19
Express, 181

Farming, 126
FIDISOL, 20, 101
Finite difference method, 16, 139
Finite element method, 16, 51, 52, 139
Fortran, 9, 11, 12, 20, 23, 24, 25, 26, 29, 33, 34, 101, 104, 126, 131, 134, 136, 178, 179-182, 189, 208, 229
Fujitsu VP, 2, 19

Gather operation, 8, 14
Gauss elimination, 14, 62, 63, 143, 144
Gauss quadrature, 43, 45
Gauss-Chebychev integration, 44
Gauss-Legendre quadrature, 44
Gauss-Seidel method, 63, 142, 143
Gram-Schmidt orthonormalization, 85
Grand canonical ensemble, 211, 216
Green's function, 162